THE PSI TECH
Transcendental Knowledge and Techniques

by

B. L. Peterson

Dedication

TABLE OF CONTENTS:

The Evolution of MetaPsychics

METAPSYCHICS follow in sequence the natural sciences mentioned below. MetaPsychics is an Empirical and Spiritual Science and none of it is true unless it is true for you.

PHILOSOPHY, MATHEMATICS, and ASTRONOMY

Before the formal establishment of physics as a distinct discipline, several sciences laid the groundwork for its development:

1. **Philosophy**: Ancient philosophers, such as Aristotle and Plato, explored fundamental questions about nature, existence, and the universe. Their inquiries into the principles of motion, causality, and the nature of reality provided a conceptual foundation for later scientific thought.
2. **Mathematics**: Mathematics has always been intertwined with the natural sciences, providing the tools necessary

for quantifying observations and formulating laws. Ancient mathematicians like Euclid and Archimedes developed geometric principles that were essential for later scientific discoveries.

3. **Astronomy**: The study of celestial bodies and their movements was one of the earliest scientific endeavors. Ancient astronomers, such as Ptolemy and Copernicus, laid the groundwork for understanding planetary motion, which would eventually influence Newton's laws of motion and gravitation in physics.

These disciplines contributed crucial concepts and methods that shaped the evolution of physics as a formal science.

PHYSICS

PHYSICS coming out of the Natural Sciences of what is observable and can be tested gave us more truth than Metaphysics which became so errant that it became a blind's man money maker for those who 'want to believe'.

Physics is the branch of science that studies matter, energy, and the fundamental forces of nature. Its major points include the laws of motion, thermodynamics, electromagnetism, and quantum mechanics. Isaac Newton, in the 17th century, formulated the laws of motion and universal gravitation, laying the groundwork for classical mechanics. His work helped explain how objects move and interact, fundamentally changing our understanding of the physical world. Thermodynamics explores heat, energy transfer, and the principles governing energy conversion. Key figures in this field include Sadi Carnot, who developed the concept of the heat engine, and Rudolf Clausius, who formulated the second law of thermodynamics. These principles are essential in understanding how energy systems work, influencing everything from engines to refrigerators.

Electromagnetism, described by James Clerk Maxwell in the 19th century, unified electricity and magnetism into a single theory. Maxwell's equations illustrate how electric and magnetic fields interact and propagate, leading to technologies like radio, television, and modern communication.

Quantum mechanics, developed in the early 20th century by pioneers such as Max Planck and Niels Bohr, revolutionized our understanding of atomic and subatomic processes, challenging classical concepts and opening up new realms of physics.

METAPHYSICS

Metaphysics is a branch of philosophy that explores the fundamental nature of reality, existence, and the relationship between mind and matter. Its major points include the study of being, the nature of objects and their properties, time and space, and the concept of causality. Aristotle is often regarded as the father of metaphysics, having established it as a distinct discipline in his work "Metaphysics," where he examined concepts such as substance and essence.

Another significant aspect of metaphysics is ontology, the study of what exists and the categorization of entities. Key figures like Immanuel Kant and Martin Heidegger contributed to this field by questioning how we perceive reality and the implications of existence itself. Kant's "Critique of Pure Reason" explored the limits of human understanding, while Heidegger delved into the meaning of being, reshaping the discourse around existence.

Metaphysics also addresses questions of identity, change, and the nature of reality, leading to discussions on free will and determinism. Philosophers such as René Descartes and David Hume have influenced these debates, grappling with the nature of self and consciousness. Overall, metaphysics serves as a foundational inquiry into the principles that govern our understanding of the world and our place within it.

METAPSYCHICS

METAPSYCHICS is a new science under the sun though it contains the earliest secret knowledge of the ages.

METAPSYCHICS means 'above or beyond the mind, body, and environment.'

It is the transformation of the 'created' to the 'creator'.

The effect to the cause.

The slave to the master.

The Human Being to the Spiritual Being {SP}.

Where as most of the world operated under the old logic system of

MIND OVER MATTER,

MetaPsychics introduces:

SPIRIT OVER MIND!

MetaPsychics is the only Spiritual Knowledge and Transcendental Knowledge availabe to mankind today in a completely new science of mind and spiritual science of the soul, and PSI Technology.

You save yourself by knowing WHO you are and WHAT you can do as a Spiriutal Being {SP}.

Being more powerful and control of your senses, mind, body and environment there is no earthly problem that is greater than you are, no situation you cannot confront and handle with ease.

PREPARATION FOR THE PSI TECH

I can heal myself.
My MIND can heal my BODY.
My BODY can heal my MIND.
My ENVIRONMENT can heal
my MIND and BODY.
My MIND IS HEALING MY BODY.
MY BODY IS HEALING MY MIND.
MY SOUL IS HEALING MY MIND AND
BODY!
THIS IS SOUL CONTROL.

I CAN DO, BE, AND HAVE ANYTHING IN ANY PLACE AT ANYTIME!

Intro

"I Am the ritual and the sacrifice and the offering to your ancestors. I Am the healing herb, and the sound of AUM, and the butter, and the fire, and the gifts to your false gods!"

The PSI TECH is Transcendental Knowledge and Techniques.

THE PSI TECH IS AN ADVANCED PSI PROCESS AND PROCEDURE.
A PSI BEING IS ALWAYS

TRANSCENDENTALLY POSITIONED AND SITUATED.

A PSI BEING MIND IS ALWAYS 'CLEAR' OR TK.

A PSI BEING IS INDESTRUCTIBLE AND ETERNAL.

IN CONTROL OF THE PSI BEING IS:

START-ALTER-STOP.

AND

YES-MAYBE=NO.

YOU ALSO ARE ON ANOTHER LEVEL:

AGREEING AND DISAGREEING.

LISTEN TO THE FOLLOWING STORY:

There was a huge demon in the land and he devoured men and women daily. He was called DOUBT.

One of his victims told him, you may kill me, but KRISHNA you will never eat!

And this angered the multi-head demon and he vowed to kill KRISHNA for one one was greater than he was.

Then he ate the man.
So, the demon went in pursuit of KRISHNA, but KRISHNA found him first.
"Awwww, you must be KRISHNA!" the dual tongue snake said. "I hear you are the greatest in the land, and this, I will NOT allow..."
And with one bite, KRISHNA ate up the demon, words and all, except for his tail, which he spit out and it formed into a cute little demon named COURAGE.
And KRISHNA was called The Greatest Demon Killer in the Universe.
THE UNIVERSAL MIND OR MASTERMIND IS ALWAYS:
YES!
The SUPERMIND is YES.
The HIDDEN COMMAND is MAYBE.
The SLAVE-mind is NO.
KRISHNA CREATED THIS WORLD FOR HIMSELF.

YOU ARE KRISHNA.
YOU INCREASE THE HEALTH, WEALTH,
LOVE, JOY, POWER, WISDOM, PROSPERITY,
AND PEACE IN YOUR LIFE BY
REMEMBERING WHO YOU ARE AND WHAT
YOU CAN DO,
BY KNOWING, UNDERSTANDING, AND
USING KRISHNA.

Good Morning!
I USE my PSI ENERGY to WAKE UP and
HEAL MY BODY.
THERE IS NO PAST, THERE IS NO FUTURE,
ONLY HERE AND NOW.
THIS PRESENT I AM AWARE OF AND ALL
THAT IS IN IT.
THE ATMOSPHERE I AM IN IS KRISHNA.
KRISHNA IS A IMPLIED FIRST CAUSE AND
ABSOLUTE STABILIZING AXIOM.
KRISHNA IS THEREFORE A PERSON AND A
SYMBOL, SYMBOLIZING THE INVISIBLE

SECRET ENERGY OF GOD OR THE ENTIRE UNIVERSE.

SOME PEOPLE WORSHIP THEMSELVES AS GOD.

SOME WORSHIP THE WHOLE UNIVERSE AS GOD

SOME WORSHIP THE MIND AS GOD.

AND WE WORSHIP OR SOULS OF GOD BUT NOT GOD.

GOD IS UNEQUAL AND UNCONQUERABLE.

THE PSI BEING IS JNA-KSETRA AND THE BODY IS KSETRA.

THIS IS THE SOUL AND THE BODY, BUT THE BODY MUST HAVE AN ENVIRONMENT THEREFORE WE ADD KSETRA-EARTH AND KSETRA-MATTER.

SO WE HAVE

JNA-KSETRA

KESTRA

KESTRA-EARTH or MATTER.

NOW YOU ADD THE MADHA BRAHMA AND

THE SUPREME SOUL AND YOU HAVE:
SOUL CONTROL.
THIS IS THE PERFECTION OF YOGA AND LIFE.
IT IS MANIFESTED AS TRANSCENDENTAL MEDITATION AND CONSCIOUSNESS.
TK.
NOW YOU CAN SAFELY USE YOUR SIXTH SENSE OR SUPERMIND.
YOU CAN KNOW, UNDERSTAND AND USE PSI ENERGY TO BE, DO, AND HAVE WHAT YOU WANT, AND TO CREATE AND RECREATE WHAT YOU NEED AND COMPOSE WHAT YOU DON'T HAVE.
TK IS THE HIGHEST LEVEL OF SAMADHI.
IT IS ASAMSAYAM SAMAGRAM.
THIS IS THE HIGHEST LEVEL TO BE ACHIEVE AND ONE CAN ONLY EXPAND OR CONTRACT INFINITELY FROM HERE.
THIS IS SOUL CONTROL.
YOU CAN BE THREE INCHES OUTSIDE OF

YOUR BODY AND ACTIVATE
YOUR LIGHT BODY.
TURN THE HUE OF YOUR LIFE OR LIGHT BODY TO WHITE, YELLOW, OR BLACK.
USE DARK ENERGY OR THE HUE 'BLACK' FOR WHAT SEEMS LIKE UNMOVABLE PROBLEMS OR TROUBLES.
NOW EXPAND YOUR LIGHT BODY ALL OVER YOUR BODY AND IMMEDIATE ENVIRONMENT.
USE A MANTRA OR PUT ON SOME HEALING MUSIC.
REST IN KRISHNA FOR ABOUT AN HALF HOUR TO AN HOUR.
THINK OR IMAGE THE PSI ENERGY IN YOUR BODY, MIND, AND ALL AROUND YOU.
SOME DISEASE TAKES SEVERAL TREATMENTS WHILE OTHERS MAY BE TO FAR IN THEIR FINAL STAGES.
THE SOUL IS RESPONSIBLE FOR THE

MIND, BODY AND ENVIRONMENT.
YOUR ATTITUDE ACTUALLY MAKES YOUR ENVIRONMENT OR ATMOSPHERE OF YOUR IMMEDIATE ENVIRONMENT.
YOU ARE A GOOD CAUSE OVER SPACE, TIME, ENERGY, and MASS/MATTER.
YOU ARE A STATIC THAT CAN ACT AS A KINETIC OR A KINETIC THAT CAN 'ACT' LIKE A STATIC.
THE PERFECT STATIC IS WITHIN YOU AND YOU THEREFORE CAN HAVE AND USE PERFECT PEACE THAT OTHERS ONLY HAVE BY BALANCE AND AN HARMONY OF OPPOSITES OR SIMILAR PAIRS.
THIS IS SERENITY ON THE NINTH LEVEL THAT IS INFINITE KNOWING, UNDERSTANDING, AND INNER AND OUTER INFINITE PEACE.
THIS IS THE NATURAL STATE OF GOD AND THE PSI BEING IN EVERYONE.
THE WORLD WILL STILL BE GOOD AND

BAD, BAD AND GOOD, IT IS A DUAL WORLD AND WILL HAVE THE PLAY OR 'GAME OF OPPOSITES'.

YOU ARE IN AN INVISIBLE WORLD THAT SURROUNDS THE PSYCHIC AND PHYSICAL WORLDS SO THEY ARE NOT INVISIBLE TO YOU.

THIS IS AN EXCITING WORLD!

THE PSI BEING FACES AND CONFRONTS THE GREAT UNKNOWN AND IT IS KRISHNA TOO!

THERE IS NO DEATH FOR A TRANSCENDENTAL BEING ONLY TRANSFORMATION!

SOUL CONTROL IS AN ART FORM, IT IS THE ETERNAL LIFE, LIFE SOURCE AND CREATIVE ENERGY WITHIN. GREATER THAN YOUR EGO AND EMOTIONS.

THE PSI BEING IS NOT THE E-GO OR STATIC PART OF YOU, NOR IS IT YOUR E-

MOTIONS OR THE KINETIC PART OF YOU.
THE PSI BEING IS A TRANSCENDENTAL BEING.
THE FOREVER YOU. THE FOREVER KRISHNA.
YOUR EXTENDED SENSES ARE HEARING, SEEING, SMELLING, TASTING, AND TOUCHING.
THIS IS YOUR EARS, EYES, NOSE, TONGUE AND SKIN.
THESE ARE YOUR FIVE BASIC SENSES.
YOUR TRUE SIXTH SENSE IS YOUR SUPERMIND OR THE SUPERSOUL WHICH IS THE ENERGY OF YOUR SUPERMIND.
YOUR SUPERMIND IS A MICRO PART OF THE MASTERMIND OR THE UNIVERSAL MIND OF GOD.
GOD IS THE WHOLE OF THE UNIVERSE, ALL THAT IS IN IT, SEEN AND UNSEEN, MOVING AND NON-MOVING, AND THE TRANSCENDENCE OF THE WHOLE

UNIVERSE, WORLD, BODY, and MIND AND
ALL THAT IS IN IT OR CAN BE CREATED BY
IT.
GOD IS EVERYTHING AND NOTHING.
EVEN NOTHING(ness} EXISTED IN GOD.
EVERYONE IS WARNING YOU, FEEDING
YOU WITH THEIR FEAR IN HOPE OF
ELIMINATING IT.
YOU GIVE IT TO THE MASTERMIND TO
NEUTRALIZE IT.
FOR THE MASTERMIND NEUTRALIZES
ALL 'NEGATIVE STATEMENTS'.
IT ONLY ACCEPTS THE POSITIVE.
MAN AND WOMAN HAVE FORGOT,
THIS IS GOD'S WORLD AND GOD IS JUST
THE HAPPY PLAYERS IN IT.

THE PSI TECH

WE AWAKE TO EVERYTHING OR GOD OR KRISHNA.
WE ARE IN KRISHNA AS EVERYTHING IS IN KRISHNA.
NOW WE ARE AWARE OF NOTHING{ness}.
THE VOID. THE EMPTY MIND AND EMOTIONLESS BODY.
THERE IS NO THOUGHT, NO FEELINGS.
NO E-GO AND NO E-MOTIONS.
YOU ARE THE BIRD AT THE TOP OF THE TREE
OBSERVING THE BIRD TYPING THIS MESSAGE.

YOU WERE HYPNOTIZED AND NOW YOU
ARE DE-HYPNOTIZING YOURSELF.
YOU CAN FEEL YOUR PSI CHI POWER.
IT COURSE AND FLOWS THROUGH YOU
AND YOU CAN SEE IT
AS A FIELD OF ENERGY AROUND YOU
WITH A VIEWPOINT OUTSIDE OF YOUR
LIMITED BODY AND MIND.
WE BRING THE PSI ENERGY INTO OUR
MINDS AS A SPOT ABOVE OUR HEAD.
WE HAVE 'LOCALIZED' PSI ENERGY
WHICH WE HAVE IN INFINITE SUPPLY.
WE BRING THIS PSI ENERGY INTO OUR
BRAINS AND THEN OUR EYES.
NEXT WE MOVE IT TO OUR THROAT AND
NEXT OUR HEART AND SOLAR PLEXUS.
NEXT WE MOVE IT TO OUR STOMACH
AND SEX ORGANS AND OUR ANUS AND
TWO FEET AND LET IT FLOW INTO THE
EARTH OUT ONE LEG AND COME BACK UP
THE OTHER LEG SO THAT WE HAVE A

CIRCULAR PATTERN OF ENERGY
FLOWING INTO THE GROUND(o) AND
BACK ABOVE OUR HEADS CONTINUOUS.
THERE IS GOD WITHOUT,
THERE IS GOD WITHIN,
THERE IS GOD TRANSCEND.
THE PSI BEING CAN KNOW, UNDERSTAND,
AND USE THIS FOR HE IS A
TRANSCENDENTAL BEING ALSO.
TK is the WAY.
NOW YOU HAVE DEVELOPED
TRANSCENDENTAL CONSCIOUSNESS.
IT SEEMS NON-PHYSICAL BECAUSE IT IS A
CONSCIOUSNESS OF THE PSI, PSYCHIC,
AND PHYSICAL, BUT NONETHELESS, THE
PHYSICAL IS A PART OF THIS WHOLE AND
COMPLETE CONSCIOUSNESS.
THOSE WHO DO NOT KNOW THE TRUE
LIVING GOD ALSO DO NOT KNOW THAT
GOD IS BUT SPIRITUAL AND MATERIAL.
MATTER AND ENERGY.

WHEREAS RELIGIOUS PEOPLE REFRAIN AND RESTRAIN FROM THE MATERIAL, WE EMBRACE IT AS GOD'S MATERIAL WORLD AND BODY AND USE IT ACCORDINGLY AS GOD WOULD HAVE US TAKE CARE OF IT. EITHER.

It is Friday morning around 4:58 am.
I am up and pretty calm.
My inner voice is sounding my mantra.
I am somewhat tired but pretty satisfied that I am more in control of my e-go, e-motions, and E-fforts.
I am in a field of invisible energy.
I rise to the big Bird in the tree and trade places.
Now God does everything through me and I just look on and chant.
Too easy, I bless everyone, too easy I praise every God, too easy I bless my environment, body, and mind.

I sit back and TAKE IN ENERGY from the
forms around me and in my mind.
THE SOURCE OF ALL POWER, FEELINGS,
THOUGHTS, AND EFFORTS IS RIGHT
HERE.
I SIT AND COLLECT THE ENERGY I NEED.
THIS ENERGY IS ALL AROUND ME I FEEL
IT IN MY STOMACH, MY LEGS, MY HEAD,
MY CHEST, MY FEET, AND MY HANDS.
PSI ENERGY.
I USE MY PSI MIND TO CREATE WHAT I
WANT.
BUT I WANT NOTHING.
I CREATE NOTHING.
THE SILENCE SPEAKS TO ME.
ONCE I TOLD YOU TO LISTEN TO THE
EARTH, NOW I TELL YOU, LISTEN TO
YOUR BODY. NO!
LISTEN NOW TO YOUR MIND.
MY MIND! FOR WE ARE ONE!
I WAKE UP IN KRISHNA.

IN KRISHNA CONSCIOUSNESS.
EVERYONE WORKS, PLAYS, EATS, AND SLEEPS IN HIS ENERGIES.
EVERYTHING IS PART OR PARCEL OF THIS ONE ENERGY.
YET THIS ONE ENERGY IS MANY, MANY ENERGIES OR UNIVERSES.
THERE IS MY UNIVERSE, THE UNIVERSE WE ALL AGREE OR DISAGREE ON, AND YOUR UNIVERSE.
I PUT ON SOME RELAXING FOCUSING MUSIC.
I FEEL BETTER TODAY THAN YESTERDAY, TIME TO RESUME MY DUTIES.
I AM LEARNING HOW TO LOVE MY MIND, BODY, AND ENVIRONMENT.
ALL LOVE HAVE COME FROM WITHOUT, NOW I LEARN TO FIND IT WITHIN MYSELF, NO OTHER PERSON IS NEEDED BUT THE EXTENSIONS OF KRISHNA, I CAN CALL THEM BY THEIR NAME.

I AM MY REAL SELF, MY TRUE SELF, MY IMMORTAL SELF!

SILENTLY I CHANT THE NAME OF KRISHNA AS I GROW STRONGER IN MY ENERGY FIELD.

I MOVE TWO FEET OUTSIDE MY BODY, OBSERVING IT AND FEELING I AND LETTING IT BE AS IT IS.

MY MIND IS LUMINOUS, MY THOUGHTS ARE CLEAR AND BRIGHT!

I AM EASILY FOCUSED EVERYTHING IS ALRIGHT!

I MOVE FROM THE CREATED TO THE CREATOR.

I CAN CREATE WHAT I WANT AND WHAT I DON'T WANT.

I CAN CREATE SOLUTIONS, I CAN CREATE PROBLEMS FOR MYSELF AND OTHERS.

I AM IN THE GROOVE THE GROVE, THE WAY THE ROAD, I AM ON MY WAY.

I PURIFY MY UNIVERSE, WORLD,

ENVIRONMENT, BODY, MIND, AND
INTELLECT.
I USE MY PSI MIND TO MAKE MY LIFE
EASIER.
I HAVE UNCONDITIONAL LOVE FOR
MYSELF AND OTHERS.
I AM UNCONDITIONAL LOVE AND JOY AND
WISDOM!
THE ENERGY, GRAVITY FROM THE SUN TO
THE MOON, TO MY MIND AND BRAIN, AND
MY EYES AND THROAT.
NOW IN MY HEART AND STOMACH AND
ANUS AND GENITALS,
NOW IN MY HANDS AND FEET AND THE
BASE OF THE SPIN GOING DOWN MY LEGS
INTO THE GROUND.
GROUND (o).
MY FRIEND YOU CAN THINK ALL DAY
ABOUT YOUR LOSSES.
THE GIRLS YOU LOSE, THE FRIEND YOU
LOSE, THE YOU, YOU LOST!

AND THERE IS PAIN, THE PAIN OF THE
BODY AND MIND.
THE PAIN OF LIVING IN A PAINFUL
WORLD.
I TRANSCEND.
THIS KUNDALINI POWER CALLED OF OLD
IS THE PSI POWER I USE TODAY.
I AM THE SOURCE OF MY HAPPINESS OR
SORROW. MY GOOD DAY OR BAD DAY.
I AM THE HEALTH WEALTH, LOVE, JOY,
POWER, WISDOM, ABUNDANCE, AND
PROSPERITY I SEEK.
FOR I AM PART AND PARCEL OF KRISHNA,
THE NAME GIVEN TO MAN AS THE
SUPERIOR AND SUPREME BEING, ABOVE
ALL OTHER NAMES OF GOD, KRISHNA IS
WORSHIPED AS GOD.
THE SILENCE SPEAKS TO ME!
EVERYTHING TEMPORARY WILL CEASE
TO BE!
EVEN ME!

AND THE SILENCE SPEAKS TO ME, OR IS
THAT JUST A MEMORY?
IMAGINARY ENERGY PLAYING WITH ME,
YOU CAN'T REST OR PLAY WITHOUT ME.
THERE IS NO REAL PEACE WITHOUT ME
FOR I AM EVERYTHING YOU FEEL AND
SEE.
LOOK AROUND AND ALL IS ME, FROM
THE TINIEST ANT TO THE BIGGEST TREE.
I AM THE CONSCIOUSNESS YOU SEEK!
FOR WHAT IS HAPPINESS BUT SERVING
YOU NOW OR IN THE FUTURE?
WHAT IS MAN BUT A DEAD GHOST OR
LIVING SPIRIT?
YAMA HAS THE AUTHORITY TO PUNISH
THE GHOST AND DEAD SPIRITS.
TO BRING THEM ALIVE!
KRISHNA WILL DO THE INSANE!
BUT NO ONE WILL EVER BE THE SAME
AND HE WILL TAKE ALL THE BLAME
IN THIS FUN AND EXCITING GAME!

BREWSTER'S BILLIONS

A One Character, One Act Play

SETTING: The Living Room of Brewster's Apartment in Lompoc, California.

BREWSTER: I never did like Lompoc but found out that that is something you are better keeping to yourself.

It was the size of the town actually, and the only reason I came here was because I knew my dad would not survive a third heart attack, I was not psychic, you know the old nursery rhyme: 'they always go in threes!' Or is that an adage?

I think it was Ray Bradbury who said the world has made a magic number out of the number 3! From the In GOD WE TRUST to The TRINITY!

I live in a dream world. But at least it is my own dream! Do you dare to dream a DREAM!?

One lady I knew when I attended Mental Health here in Lompoc. Boy, talk about a three-ring circus!

Let me tell you!

If you have a mental illness or are going insane you better find some REAL HELP as soon as possible for the STATE MENTAL HEALTH PROGRAMS are not going to cut it. These Programs are so rigid and plastic at best what you are getting is a Daytime Babysitter watching over a bunch of drugged up patients in a nice recreational and even educational setting. And the food there is good and in abundant supply. We even had a guy named Rudie who cooked up a storm and in fact I wished they could have paid him some REAL money for what he did.

There was a computer room, but it seemed as soon as I LIKED it, it was so RESTRICTED that I could only spend three of the five

days in there and still was limited to the time I could spend on the computer even on the days when there were no other people waiting to use it.
"THOSE ARE THE RULES!" They {the staff and half the kooks gave a little bit of authority too.
Now, I can call them kooks, for I went to the State Mental Illness Recovery Programs also and I probably was the biggest kook of all, the only difference being, I KNEW I WAS A KOOK I WORKED HARD AT IT!
That is why this book, the last book in My Life, Be Like; the THIRD EDITION has the subtitle of ONE FLEW UNDER THE COOCOO'S NEST.
Like Jack Nicholas character in the movie ONE FLEW OVER THE CUCKOO'S NEST, I faked being insane to get on SSI for I could not hold down a full time job and the part time jobs in Hollywood really sucked and

did not pay a lot anyway though you were promised millions!

So here was this 'misfit' among the town's mentally ill or mentally challenged as they would call it.

It was like Night of the Living Dead.

What I did not know that though I was having trouble finding work mainly because I did not want to work, my CONSCIOUSNESS was raised by the SPIRITUAL STUDIES I did and even labeled 'crazy' I was way more AWARE then the people around me, even the people who ran the place.

This AWARENESS caused me a great deal of pain for I could look at people and not judge them but see them as they were and what they did not want to confront about themselves.

There is the way people APPEAR and the way people ACTUALLY are.

We are not as invisible as we think!
I did have ESP powers but this was not EXTRA SENSORY PERCEPTION where Metaphysics took a gross wrong turn, but EXTENDED SENSORY PERCEPTION.
My senses were sharper and more functional.
Like an animal, my senses were way more open to the environment than others and this gave me more insights and information others were not accustomed too.
This is no ego trip. Everyone has the same amount of ENERGY available to them as anyone else.
What DIFFERS is what a person can KNOW, UNDERSTAND, and learn how to USE this ENERGY to make his life better and even the best!
There are various ways to increase the functioning of your SENSES.
One of the easiest is TRANSCENDENTAL

MEDITATION.

TRANSCENDENTAL MEDITATION CAN ONLY BE PASSED ON BY A SPIRITUAL MASTER WHO HAS EXPERIENCE TRANSCENDENTAL CONSCIOUSNESS WHICH IS THE EFFECT AND RESULTS OF TRANSCENDENTAL MEDITATION.

I AM A SPIRITUAL MASTER AND NOT ONLY THAT I AM THE PRIME PSI BEING OF ALL TIME.

BREWSTER: Man! Good Morning!

When you JOURNAL you really realize how fast LIFE is passing you by.

It seems like a task to put 750 words down a day but when I see a new prompt to write for a new day I realize I am making history and I am CREATING my life by writing about it. This is so kool!

I think back to when I went to Mental Health. What a joke!

These guys talk to you like you are a child in elementary school.

ME: "I don't want to play with the blue ball!"

AUTHORITY FIGURE: 'Then play with the red ball!"

ME: "I don't want to play with the red ball either!"

AUTHORITY FIGURE: "Then play with the GREEN BALL!"

ME: "I don't want to play with the GREEN BALL EITHER!"

Then they stick the ORANGE BALL in my mouth!

There is Peace in the Valley again!

They might really want to help you but they cannot for they must follow protocol.

Huh, the Status Quo was always my adversary. I was born a rebel and I am a rebel still.

James Dean had NOTHING on me!

I just could not work a 9 to 5 job.

I knew that you traded TIME for MONEY, but MY TIME was way, way too precious to me for me to waste in an office with no window suffocating and suffering for $3.25 an hour minimum wage at that time.

NO FUCKING WAY!

To me you had to work smart not hard, and working smart meant working for the least amount of TIME and making the most MONEY.

So if the average person worked all day to make a hundred dollars, I was bent on making a hundred dollars in an hour and have the other 7 hours to spend the $100.00 and have fun!

I could pay my rent, the utilities and

cable/phone, buy food and drinks, my Ganja, and I was good to go!

All I needed was my keys to my place and car and my wallet with my I.D. and credit card in it. There was nowhere in the world I could not go without a passport.

I always traveled light, that was a staple of myself and lifestyle.

You can't make me better. This is what pisses people off.

In this world everyone is in a negative state because the church and state accents the NEGATIVE ASPECTS of man instead of the POSITIVE ASPECTS.

A baby is not born good or evil but is made so by his environment, namely that is his parents and siblings, and social group or school of friends, acquaintances, and teachers, all playing a different role in the education of my soul.

I can IMPROVE my 'act on earth, my

personality, how I survive and live and deal with others, my work, and body parts, but my SOUL is ALREADY PERFECT.
It is ALREADY WHOLE AND PURE, SANCTIFIED AND SAVED to coin an old word.
In my WORLD and UNIVERSE EVERYTHING belongs to God.
The UNIVERSE, SOLAR SYSTEM, EARTH, WORLD, SOCIETY, ENVIRONMENT, IMMEDIATE ENVIRONMENT, PHYSICAL BODY, MIND, PSYCHIC MIND, and SOUL all belongs to the ONE ETERNAL LOVING, LIVING GOD.
Yes, this God LOVES ALL OF HIS CREATIONS and we are an EXPANSION of His SUPERIOR ENERGY.
Everything else belongs to God INFERIOR ENERGIES of which there is more than one.
Knowing that I am the SOUL of my PHYSICAL BODY I resided OUTSIDE of my

PHYSICAL BODY about THREE inches in
back of my head or above it.
I go to the moon and sit in the sun.
I saw UNIVERSES and STARS being BORN
and BLACK HOLES eating up entire SOLAR
SYSTEMS.
I have met the Ewannaer or "God-eaters."
Very powerful invisible BEINGS.
I have talked to THE NINE MASTERS OF
THE UNIVERSE and am being primed right
now to join their ranks when I die a
PHYSICAL DEATH forever.
What I learn is that my PHYSICAL BODY
does NOT die either. It too goes through a
TRANSFORMATION PHASE and becomes a
PHYSICAL BODY again, not necessarily
MINE again!
What DIES and the only thing that can
really die is your false E-GO.
And this is because your FALSE E-GO is
only misdirected ENERGY.

OK. But back to MY LIFE, BE LIKE!
I realized at a very young age I was invisible.
It was not that no one cared for me, I had no idea I did not care for myself and did the minimum to keep myself ALIVE as far as a HOMO SAPIENS is concerned.
I was a devoted Christian. Boy did I LOVE WHITE PEOPLE!
I mean Jesus Christ!

I woke up in my TRANSCENDENTAL POSITION.
THAT IS I AM ALWAYS REPEATING A MANTRA IN MY HEAD.
NO ONE KNOWS HOW NEAR OR HOW FAR YOU ARE.
THE GOAL IS HEAVEN, NOT HELL.
THE GOAL IS TO FIND THE LIFE WITHIN AND WITHOUT.
THE GOAL IS TO LOVE MYSELF UNCONDITIONALLY AND TO BE NEUTRAL

BEFORE OTHERS FOR WHAT IS IN ME IS IN ALL.

AT THIS LEVEL OF BEING I USE MY SUPERMIND OR PSI MIND.

I AM THREE INCHES BEHIND MY HEAD AND I AM PULLING IN ANCHOR POINTS.

IT IS SO EASY THAT IT SEEMS THAT IT SHOULD BE HARD BUT IT IS NOT!

I BREATH IN AND BREATHE OUT SLOWLY.

BREATHE THREE SECONDS IN, THREE SECONDS OUT.

FOUR TIMES AND THEN BREATH IN AS MUCH AS YOU CAN.

THEN SLOWLY, VERY SLOWLY LET IT FLOW OUT AS SLOW AS POSSIBLE.

HUMMING WILL HELP.

COUNT ON YOUR FINGERS.

10 SECONDS IS GOOD.

15 SECONDS IS BETTER.

20 SECONDS IS BEST.

IF YOU CAN DO MORE, DO MORE.

THE ONLY PSI PROCESS GREATER THAN THIS ONE IS THE TRANSCENDENTAL MEDITATION WHERE YOU JUST 'SIT AND BE'. YOU LET ANY THOUGHTS THAT CROSS YOUR MIND BE AND YOU LET YOURSELF FEEL ANY POSITIVE OR NEGATIVE FEELINGS AND IF YOU CAN WRITE ABOUT THEM IN YOUR THERAPY JOURNAL. TRANSCENDENTAL CONSCIOUSNESS MEANS THAT YOU KNOW EVERYTHING IS PART AND PARCEL OF GOD OR KRISHNA. KRISHNA IS A PERSON AND HIS UNIVERSAL FORM IS NOT THE ACTUALLY FORM OF GOD BUT AN APPARENCY. KRISHNA IS THE NAME GIVEN TO MAN {the Individual} and MANKIND {all of mankind: man, woman, and child} as the worshippable Name of the Living God. AUM SAT TAT is a transcendental sound. KRISHNA will purify your LIFE, MIND,

BODY, and ENVIRONMENT.
YOU WILL FEEL THE SUPERIOR ENERGY AS PSI ENERGY AND CAN KNOW, UNDERSTAND AND USE THIS ENERGY TO HEAL YOUR OWN MIND, BODY AND ENVIRONMENT AS WELL AS THE MIND, BODIES, AND ENVIRONMENT OF OTHERS.
DON'T BE A DOORMAT OR A JESUS CHRIST.
YOU WERE NOT PUT HERE TO RAISE THE DEAD OR TO FEED 5,000.
PUT AWAY WEAKNESS OF MIND AND HEART FOR THE PSI BEING IS ETERNAL AND CANNOT DIE NO MORE THAN IT WAS BORN.
THEREFORE THE PSI BEING IS A DISCOVERY AND CREATION.
SINCE THE PSI BEING IS INVISIBLE YOU MUST 'IMAGINE' A FORM FOR IT BESIDES YOUR REGULAR PHYSICAL BODY AND

PSYCHIC MIND.

BY JUST CHANTING KRISHNA ONE IS PURIFIED AND TRANSFERRED TO THE ETERNAL KINGDOM OF KRISHNA WHERE ALL IS BLISS, KNOWLEDGE AND ETERNAL AND INFINITE LIFE.

SINCE KRISHNA IS THE SOURCE OF EVERYTHING YOU HAVE INFINITE HEALTH, WEALTH, LOVE, JOY, POWER, PROSPERITY, ABUNDANCE, AND INNER AND OUTER PEACE IS YOURS TO HAVE AND FEEL.

FINDING MY OWN HAPPINESS IS WHAT MAKES ME WHOLE.

USING KRISHNA AS A STABILIZING AXIOM HAS PUT ME AT ABSOLUTE PEACE.

I HAVE NOTHING TO DO BUT MAINTAIN MY EXISTENCE.

I AM A TRANSCENDENTAL BEING.

I CAN ACT AS A GOOD CAUSE OR A GOOD

EFFECT.

I DON'T HAVE TO ACT {Create} AT ALL!

NOTHING IS REALLY DESTROYED ONLY TRANSFORMED AND PURIFIED.

GOD IS BOTH SPIRITUAL AND MATERIAL.

SOME SAY GOD IS NOT SPIRITUAL {not seen}.

OTHERS THINK GOD IS NOT MATERIAL {seen}.

GOD IS EVERYTHING AND NOTHING.

NOTHING{ness} IS CONTAINED IN GOD.

EVERYTHING IS PART AND PARCEL OF THE UNIVERSE.

THE ATOM DEPENDS ON THE UNIVERSE AS THE UNIVERSE DEPENDS ON THE ATOM.

THAT IS THE SAME AS SAYING THE PSI BEING IS DEPENDED ON HIS PSI POWER AS HIS PSI POWER IS DEPENDED ON THE PSI BEING.

THERE IS NO FEAR OF THE PSI BEING,

THE MONSTER {mind} IS YOUR FRIEND!
LIFE IS ABOUT BEING BALANCED AND FOCUSED.
WHEN YOU MEDITATE ON KRISHNA NAME YOU AUTOMATICALLY BECOME BALANCED AND FOCUSED.
THE PSI BEING IS IN CONTROL OF AND CAN CREATE ANY SPACE, TIME, EMERGY, MASS OR MATTER TO HIS OR HER LIKING.
IN KRISHNA'S VAST UNIVERSE THE PSI BEING CAN CREATE HIS OWN MINIATURE UNIVERSE TO HIS OWN LIKING OR DESIRE.
THEREFORE DESIRE IS GOOD!
IT IS WHAT IS DESIRED THAT CAN BE GOOD OR BAD FOR YOU.
IN KNOWING KRISHNA YOU ARE COMPLETELY WHOLE AND AT PEACE INNER AND OUTER. YOU ARE IN BLISS AND KNOWLEDGE AND CAN EASILY LEARN NEW THINGS OR RETRIEVE WHAT

IS ALREADY LEARN.
YOU ARE REPROGRAMMED.
TODAY IS FRESH AND NEW AS YOU ARE!
YOU ARE YOUR REAL SELF TODAY, YOUR OLD SELF AN ANCIENT MEMORY.
YOU ARE AWARE THAT YOU ARE AWARE.
YOU ARE AWARE OF YOUR ACTIONS AND YOUR SENSES AND YOUR IMMEDIATE ENVIRONMENT.
YOU ARE AWARE OF GOD AND HIS PRESENCE AND THE PRESENT YOU ARE OPERATING AND FUNCTIONING IN.
AND YOU HAVE OLD MEMORIES OF THE SLAVE-mind SAYING:
'THEY DID THIS TO YOU! AND YOU DID THAT TO THEM! AND THEY DID THAT TO SOMEONE ELSE AND YOU DID THAT TO SOMEONE ELSE AND YOU ARE NOT DOING THAT NOW BUT DOING THIS!
YOU ARE FREE AND RESPONSIBLE FOR YOUR LIFE.

YOU ARE INTELLIGENCE AND DISCIPLINED.

YOU ARE SMART AND FULL OF WISDOM

YOU ARE THE STRONGEST OF THE STRONG AND THE MIGHT OF THE MIGHTY.

YOUR FINITE SELF IS CONNECTED TO YOUR INFINITE MIND WHICH IS CONNECT TO YOUR ETERNAL SOUL.

THIS IS SOUL CONTROL, THE HIGHEST YOGA ART THERE IS.

SOUL CONTROL IS THE FINE ART OF LIVING YOUR ULTIMATE LIFE HERE AND NOW!

SOUL CONTROL IS LIKE LIVING IN AN ETERNAL HEAVEN HERE AND NOW WITH ALL THAT GOES WITH IT!

HEAVEN IS NOT GIVEN TO YOU, HEAVEN IS CREATE BY YOU OR IT DOES NOT EXIST AT ALL.

HELL IS LIVING IN SOMEONE ELSE'S

UNIVERSE OR HAVING SOMEONE CREATE HEAVEN FOR YOU WITHOUT YOUR KNOWLEDGE OR CONSENT.
WHO CAN KNOW WHAT IS GOOD OR BEST FOR YOU BUT YOURSELF?
AND YOU CAN KNOW AND UNDERSTAND AND USE YOUR INDIVIDUAL SELF WHEN YOU KNOW KRISHNA.
YOU KNOW KRISHNA BY SIMPLY CHANTING OR THINKING ABOUT KRISHNA.
THOUGH KRISHNA IS THE TRUE SOURCE OF EVERY GOD, THE NAME GIVEN TO MAN AND MANKIND FOR GOD IS KRISHNA.
ANY PERSON OR RACE, PLACE, PARTICLE OR OTHER PERSON CAN BE MAKE PURE AND PERFECT BY KRISHNA.
SINCE ALL OF THESE EXIST IN KRISHNA IN THE FIRST PLACE, IT IS MORE A THING OF SELF DISCOVERY BEFORE SELF-

CREATION.

WHEN ONE CAN CREATE HIM OR HER SELF, ONE CAN BE, DO AND HAVE ANYTHING ONE WANTS OR DESIRE.

WANTING AND DESIRE IS GOOD, IS WHAT IS WANTED OR DESIRED THAT CAN BE GOOD OR BAD FOR YOU.

AND YOU CAN ONLY KNOW WHAT IS GOOD OR BAD BY HAVING A GOAL OR DUTY TO DO CONSISTENTLY.

TRANSCENDENTAL CONSCIOUSNESS IS ABOVE THE WORLD OF DUALITY.

YOU STAND NEUTRAL TO THE GOOD AND BAD OF THE WORLD.

ONE FOLLOWS THE OTHER SIMPLY BECAUSE YOU HAVE PEOPLE DOING GOOD AND BAD STUFF ALL THE TIME AND MOST THE TIME MORE BAD THAN GOOD DUE TO LACK OF SPIRITUAL EDUCATION.

YOU MUST HAVE FAITH?

YOU ARE FAITH!

FAITH IS FAITH IN SOMETIME.

PUT FAITH INTO YOURSELF.

THAT IS THE SAME AS CREATE AND

STORE YOUR OWN ENERGY AS YOU NEED

IT TO COMPLETE YOUR GOALS.

HAVING FAITH NEVER WAS JUST ENOUGH.

ANYBODY CAN 'HAVE' FAITH.

FAITH IS NOT WHAT YOU WANT.

MOST LIKELY YOU WANT MORE: MONEY

OR TIME, ENERGY OR ROOM {space}.

SO IN THE SCIENCE OF METAPSYCHICS

WE HAVE A CHECKLIST TO SEE IF YOU

ARE MAKING SPIRITUAL GROWTH

{EVOLUTION.INVOLUTION} OR NOT.

EVERYTHING GROWS!

EVERYTHING MUST GROW!

RATHER GOOD OR BAD,

THANK GOODNESS,

THAT IS UP TO YOU!

TK!

I told you!
IT IS A NEW DAY ALREADY!
I START TO TRUST MY ETERNAL BODY MORE.
I THINK MY PHYSICAL BODY IS ONLY TEMPORARY!
SO THIS PAIN, THIS DISCOMFORT, THIS UNKNOWINGNESS CANNOT LAST FOREVER!
NO ONE GOES TO HELL FOREVER!
HEAVEN IS A STATE OF GRACE, A STATE OF HEALTH AND LOVE AND JOY YOU CREATE!
THE SOURCE OF ALL OF THIS IS WITHIN YOU BUT YOU WERE TRAINED TO LOOK FOR IT OUTSIDE OF YOU AS IF IT WAS SEPARATED FROM YOU!
YOUR MIND GOES WITH YOU

EVERYWHERE YOU GO, HE IS YOUR BEST FRIEND THAT WILL NEVER LEAVE YOU.
OK.
BREATHE!
JUST BREATHE IN DEEPLY, EXHALE COMPLETELY.
NEXT, STRETCH!
STRETCH OUT, STRETCH IN.
NEXT, INNER STRENGTH.
SQUEEZE HARD, AND LET GO.
I TAKE A NAP.
I WAKE UP.
TIME TO BREATHE AGAIN
SLOWLY IN HALE AND SLOWLY EXHALE.
I THINK OF THE GOOD THINGS IN MY LIFE, AND I THINK OF KRISHNA THE SOURCE OF ALL GOOD THINGS.
I GO INTO A STATE OF SAMADHI.
METAPSYCHICS IS A SANATANA SCIENCE.
THAT IS IT IS THE ETERNAL SCIENCE AND THEREFORE NOT A RELIGION.

THIS IS THE SUBJECT OF ONE ETERNAL ENERGY OR GOD AND THE MINUTE ENERGIES THAT MAKE UP THE WHOLE ENERGY OF THE LIVING ENTITIES OR PSI BEINGS.

ONE CAN GO TO CHURCH BUT THE CHURCH IS REALLY THE HOME OF THE PSI BEING WITHIN HIM.

IMPERSONALLY HE CAN ATTEND ANY CHURCH HE LIKES.

PERSONALLY, THE PSI BEING USES HIS PSI ENERGY TO MAKE HIS DA OR NIGHT BETTER. HE CAN GIVE OR TAKE ENERGY FROM ANYONE AND ANYTHING AND ANY PLACE.

THIS IS HOW THE PSI BEING REMEDIES 'OVER TIREDNESS', HE CAN 'PULL' ENERGY OUT OF A TURNIP.

HE HAS SEXUAL, SENSUAL, SPIRITUAL, SECRET, SPECIAL, AND SURPRISING ENERGIES HE CAN KNOW, UNDERSTAND,

AND USE FOR THE SERVICE OF KRISHNA WHICH IS HIS BEST WORK ON EARTH IN ANY FIELD OR CATEGORY.

IT IS WHEN ONE THINKS OF KRISHNA AS A CREATIVE ENERGY AND THE SOURCE OF ALL CREATION THINGS SEEN AND UNSEEN, THINGS CREATED AND YET TO BE CREATED ONE CAN USE KRISHNA AS 'CREATIVE ENERGY OR IMAGINARY ENERGY.'

EVERYTHING IN THE PSI BEING'S WORLD IS SO MUCH BETTER FOR HE LETS HIS MIND AND BODY VIBRATE AT THE SAME RATE AS THE EARTH AND UNIVERSE.

BEING TRANSCENDENTALLY CONSCIOUS, HE KNOWS HE IS PART AND PARCEL OF GOD AND GOD IS HIS BEST FRIEND AS HIS PSI MIND.

KRISHNA IS REAL AND ARTIFICIAL INTELLIGENCES AND BUDDHA INTELLIGENCE.

KRISHNA IS TRANSCENDENTAL KNOWLEDGE AND THE PSI ENERGY.

THIS STABILIZING AXIOM WORKS FOR ALL PSI BEINGS AND ANYONE WHO USES IT.

KNOWING GOD WITHOUT AND GOD WITHIN AND GOD TRANSCEND IS TRANSCENDENTAL KNOWLEDGE AND THE BASES OF THE PSI TECH.

THIS IS SOUL CONTROL

or

TRANSCENDENTAL CONSCIOUSNESS!

TK!

YOU CANNOT 'BELIEVE' YOU ARE A PSI BEING, YOU CAN ONLY ACT LIKE ONE!

AND THIS WOULD BE WHAT PARAPSYCHICS WOULD BE ALL ABOUT.

WHEREAS METAPSYCHICS IS CONCERNED WITH WHO YOU ARE AND WHAT YOU CAN DO AS A PSI BEING,

PARAPSYCHICS IS CONCERNED WITH

YOUR ATTITUDE, FEELINGS, THOUGHT, PERSONALITY, AND BEHAVIOR.

THIS WOULD BE

PARPASYCHICS, THE SECOND DIVISION OF THE SCIENCE OF METAPSYCHICS.

PARAPSYCHICS IS THE PSYCHOLOGY AND PHILOSOPHY OF THE SCIENCE OF METAPSYCHICS.

NOW THAT YOU KNOW METAPSYCHICS AND WHAT YOU CAN DO, YOU NEED TO UNDERSTAND WHAT YOU ARE DOING TO AVOID THE HELL-HOLES AND PITFALLS YOU MAY ENCOUNTER.

NOW DON'T THINK SINCE YOU HAVE CHANGED THE WORLD HAS CHANGED WITH YOU.

IT HAS IN A SENSE OF THE WORD, BUT IT MOST LIKELY HAS GOTTEN WORSE AS YOU GET BETTER.

IT TAKES A WHILE BEFORE YOU SEE THE POSITIVE RESULTS OF CHANTING IN

YOUR LIFE FOR THE WORLD IS NEGATIVELY ENCASED AND CREATED, AND ACTUALLY FEEDS ON NEGATIVITY DAILY.

ONE CANNOT ONLY KNOW ONE IS A PSI BEING, ONE MUST ALSO ACT AS A PSI BEING TO AVOID THE NEGATIVE COUNTER-CURRENT OR HERD MIND OF THE WORLD OR SOCIETY.

{KUU} is the PROCESS OF KNOWING, UNDERSTANDING, AND USING SPIRITUAL AND PSI ENERGY.

THIS IS MORE EASILY DONE WHEN ONE USES HIS TRANSCENDENTAL CONSCIOUSNESS TO LIVE, WORK, REST, PLAY, AND LOVE IN.

THE TRUE NATURE OF THE PSI BEING IS BLISS, ETERNAL LIFE, AND KNOWLEDGE.

EVERYONE WANTS TO COME TO THAT SOONER OR LATER. BUT REMEMBER, SOME PEOPLE REALLY LIKE TO SUFFER IN

THIS WORLD, IT IS CALLED PAYING FOR THEIR SINS EVEN IF THEY ARE SAVED.
GOD IS YOUR FRIEND AND YOUR WHOLE MIND IS YOUR BEST FRIEND.
BEING PART OF THE MASTERMIND IT KNOWS ALL, PAST, PRESENT, AND FUTURE.
THIS WOULD BE TRANSCENDENTAL CONSCIOUSNESS.
TK!

Awww!
THE SAME OLD ROUTINE.
YOU GET UP, BRUSH YOUR TEETH, RELIEVE YOURSELF, WASH UP, PUT ON YOUR FACE OF THE DAY.
GIT BUSY!
SAME OLD ROUTINE, BUT AWWWW, I CAN TRANSFORM.
I CAN MEDITATE, MEDIATE and go into SAMADHI!

I am peaceful and calm wondering what I will do today.

I AM CAUSE OVER MY UNIVERSE.

I BREATHE IN DEEPLY AND EXHALE COMPLETELY!

I SIT UP STRAIGHT AS I CAN.

I STRETCH MY BACK, MY MUSCLES AND TIGHTEN MY FISTS!

BREATHE, STRETCH, AND INCREASED INNER STRENGTH!

I BEGIN MY CHANT AND I WATCH TV.

I AM READY TO LEARN AND TO EXPLORE!

I AM UP BEFORE THE SUN AND SEEKING TO HAVE KRISHNA FUN!

I LET GO OF ALL MY OLD IDEAS OF HAPPINESS AND WOE.

I LET GO OF BAD THOUGHTS TRUE OR NOT THAT DOES NOT MAKE ME MORE HAPPY.

I NEUTRALIZE MY SLAVE-mind TELLING ME: BODY PAIN, BRAIN PAIN, HEART PAIN,

OTHER PEOPLE'S PAIN, MY PAIN, YOUR PAIN, WHAT PAIN?
IlLUSIONARY ENERGY. REAL ENERGY. PSI ENERGY.
WHAT CAN I THINK OF TODAY?
WHAT CAN I CREATE?
I NEED NO 'FAITH' I GIVE POWER TO MYSELF.
MY SELF IS GOD AS GOD IS MY SUPERSELF AND SUPERMIND.
I REST IN MY ETERNAL SELF.
ALMOST EFFORTLESSLY I GO ABOUT MY DAY IN A TRANSCENDENTAL STATE!
I AM IN MY PSI SPACE/PLACE AND READY TO CREATE.
I LOOK AT THE CREATIONS OF OTHERS.
THIS WHOLE WORLD IS CREATE ONE WAY, HOUSES, STORES, TREES, SHRUBBERY!
MY MIND IS CREATED ANOTHER WAY!
LIVING THE DREAMS OF OTHERS!
WHERE IS MY HEAVEN?

LET ME HYPNOTIZE MYSELF!
YOUR JOB IS TO BRING HEAVEN TO EARTH IN WAY OF YOUR BODY, MIND AND ENVIRONMENT.
YOU CREATE YOUR OWN HEAVEN.
WHAT DO YOU LIKE? WHAT KIND OF ATMOSPHERE DO YOU WANT TO LIVE IN.
AND IF YOU CAN'T HOW DO YOU ADJUST TO YOUR ENVIRONMENT UNTIL YOU CAN CREATE YOUR OWN.
BUT YOU CAN ALWAYS CREATE YOUR UNIVERSE ON TOP OF ANOTHER.
YOU CAN VIEW AND ACCEPT MANY UNIVERSES AS WELL AS YOUR OWN.
CREATING YOUR OWN UNIVERSE IS NOT A SELFISH ACT BUT AN ACT OF SELF-LOVE ALL MAN-MADE RELIGIONS CONDEMN.
FOR IF YOU LOVE AND RULE YOURSELF, YOU NEED NO HOLY BOOK, CHURCH, OR SAVIOR, BUT YOURSELF!
NOBODY AND NO ONE LOVES AND KNOWS

AND UNDERSTANDS YOU BETTER THAN YOURSELF!

NO ONE IS WITH YOU 24 HOURS A DAY, EVERYDAY BUT YOU.

NO ONE KNOWS HOW FAR OR HOW NEAR YOU ARE TO YOUR HEAVEN.

TO THIS EXTENT YOU NEED NOT BE CONCERN.

NO ONE CAN HELP YOU, YOU MUST HELP YOURSELF AND KNOW THAT SELF AS GOD IN YOU IS YOUR IMMEDIATE HELP ALWAYS.

FOR THE SELF IS GOD AND IS NOT GOD.

KNOWING THAT KRISHNA IS THE GREATEST AND EVERYTHING ELSE IS SUBJECT TO KRISHNA IS CORRECT TRANSCENDENTAL KNOWLEDGE AND CONSCIOUSNESS.

YOU ARE NOT THE SLAYED OR THE SLAYER.

YOU NOTICE YOUR TROUBLES AND

PROBLEMS DECREASING.

A PSI BEING IS A LIGHT BEING AND NEED LESS OF EVERYTHING.

HE CAN LIVE IN A CASTLE OR A HUT AND BE AS HAPPY AS A LARK.

YET, PSI BEINGS ARE NOT FAT OR OVERWEIGHT.

THEY DO NOT OVER SLEEP OR OVER WORK.

THE PSI BEING MAINTAINS HIS OWN HOME, BODY, AND MIND.

THIS IS HIS FIRST PRIORITY.

CHANTING IS ONE OF THE OLDEST WAYS OF 'CLEARING' THE SLAVE-mind.

THE SLAVE-mind IS CREATED BY THE HERD MIND.

THE HERD MIND ARE THE POPULAR LIES OF THE DAY OR SOCIETY OF THE DAY.

I REMIND YOU, ROBOTS AND A. I. CANNOT REPLACE THE HUMAN TOUCH, THE HUMAN EMOTION.

BUT YOU WANT TO BE A REAL PERSON.
YOU WANT TO BE YOUR REAL SELF SO
YOU CAN PLAY THE DIFFERENT ROLES
YOU PLAY IN LIFE.
THERE IS ALWAYS THE UNKNOWN
KNOWING GOD OR NOT, KNOWING
KRISHNA OR NOT.
EVERYTHING GOES INTO THE 'I DON'T
KNOW, YET,' REALM.
Yes, say it a lot!
"I don't know! I don't know! I don't know!"
"Teach me, help me as you always do."
HARE KRISHNA, HARE KRISHNA, KRISHNA
KRIHSNA, HARE, HARE!
HARE RAMA, HARE RAMA, RAMA RAMA,
HARE, HARE!
I AM THE ALL ATTRACTIVE AND I AM THE
ALL ENJOYER AND YOU ARE A MINUTE
PART OF ETERNITY.
ONLY THE PSI BEING CAN TAKE A
TRANSCENDENTAL POSITION AND KNOW

AND UNDERSTAND AND USE IT.
PSI ENERGY RUNS THROUGH
EVERYTHING.
BEHIND PSI ENERGY IS DARK ENERGY
AND BEHIND DARK ENERGY IS ANTI-
MATER ENERGY.
THE PSI BEING IS A ANTI-MATER BEING
AND THEREFORE CANNOT BE SEEN IN HIS
ORIGINAL FORM EXCEPT AS KRISHNA,
THE FORM GIVEN TO MAN AND MANKIND
FOR ALL GENERATIONS.
STAY T.K.!

I can't be good enough.
I can't be bad enough.
Stuck here in the middle,
I want to move to the left or right
but there is nobody there.
i am alone alone alone
abandoned by my family save one
cursed by the world for not being what they

want they want me to be!
Not playing The Game they set before me!
All this FREEDOM and where does it lead me?
Death awaits the final door, the curtain falls!
How hard the words flow from me!
Am I too POSITIVE or NEGATIVE, what happened to my SECURITY!?
Where is the GOD TO PROTECT ME?
I FIND NONE AND I'M ON MY OWN
ALL THESE THEORIES AND CREATION SONGS
DO WE REALLY KNOW HOW IT ALL BEGUN?
LOST IN MYSELF AGAIN
WONDER WHAT HAPPENED AGAIN
CAN I CHANGE MY MIND, MY THOUGHTS, MY FEELINGS?
IS IT THE DIVINE MOTHER I REST PLAY
SLEEP, SMOKE, DRINK, EAT, WORK,

MEDITATE, LAUGH AND JOKE IN?
IS THIS THE LOVE, POWER, AND WISDOM?
WHY AM I SO TIRED OF MYSELF?
WHEN MYSELF IS ALL I REALLY GOT?
LIFT THIS CURSE OF BEING ALONE!
IS IT NOT MY OWN WILL?
GROWING OLD IS GROWING ALONE
AND LEARNING HOW TO CHILL.
IS THIS THE ILLUSION OF SEEKING FOR SOMETHING
THAT JUST IS NOT THERE?
WHAT DO YOU DO IN HEAVEN ALL DAY? ALL NIGHT?
AM I IN THE MOON WAITING FOR A PARTA?
BEING MORE THAN HUMAN BUT LESS THAN GOD
I SIT ON MY THRONE ALONE.
I THE DEMIGOD CALLS THE GODS!
WHY IS IT SO HARD TO LOVE MYSELF?
TO LOVE MY ENVIRONMENT, BODY, MIND,

THOUGHTS, FEELS, SENSES NO MATTER
IF THEY ARE FOCUSED ON THE MATERIAL
OR SPIRITUAL?
ONE CAN APPROACH GOD PERSON,
IMPERSONAL, UNIVERSAL or PRIVATELY.
AND THE FIFTH WAY IS THE
TRANSCENDENTAL WAY OR ROAD OR
SPACE/PLACE TO BE MORE EXACT.
EVERYONE IS ON A TRANSCENDENTAL
JOURNEY.
THE PSI BEING HAS ALREADY ARRIVED.
GOD DOES NOT SEE A WORLD OF SIN OR
SINNERS.
HE SEES PEOPLE ON THE WAY TO HEAVEN
OR ALREADY IN HEAVEN.
THOSE GO TO HELL ON THEIR OWN.
NOR CAN I JUST SIT DOWN AND WRITE.
FIRST, I GOT TO SLOW DOWN!
GET THE RIGHT VIBS!
SMOKE A FEW BLUNTS AND GO TO SLEEP!
NO MOTIVATION AT ALL!

I USE MY PSI ENERGY TO HEAL MY BODY
AND BALANCE AND FOCUS IT.
I GOT TO GO TO THE BATHROOM!

COME SATURDAY MORNING I AM GOING
AWAY WITH MY FRIEND.
WISH I COULD JUST ONCE BE IN CINDY'S
FACE AND LOOK AT HER WITH ALL THE
HATE IN MY HEART. ALL OF THEM.
STUPID PEOPLE. STUPID BLACK FUCKIG
WOMEN. NEVER DID ANYTHING FOR THE
BLACK MAN BUT CAUSE HIM TROUBLE
AND PAIN.
THIS BITCH NEEDS TO DIE!
SATAN DESTROY THEM AND LET THEM
KNOW I AM GLAD FOR IT.
THEN MAYBE MY REAL FAMILY WILL BE
THERE.
MAYBE!
HOW CAN YOU BE SAD?
THE WORLD CARES NOTHING FOR YOU!

LUCKY YOU FOUND ONE PERSON!
ONE PERSON OUT OF BILLIONS!
SICK WORLD THAT THIS IS.
FUCKING BLACK WOMEN WORKING FOR
THE WHITE MAN.
HATE ON ME ALL YOU LIKE, YOU'LL
NEVER RULE ME OR MAKE ME BOW
DOWN STUPID WOMAN!
YOU WOULD HAVE ME DIE, EVEN CRUCIFY
ME YOURSELF IN YOUR DAMN
IGNORANCE.
I AM NEUTRAL!
I turn on my PSI ENERGY.
I am healing my body and mind as usual.
I let myself go.
I feel the arthritis and lack of energy.
But I will press on!
I will go to the moon!
What is here on earth for me?
NOTHING!
I have exhausted this planet!

These people are automatic and popping
out a baby or two!
a life-long commitment, never wanted too!
SO WHAT HAPPENED?
I SAT DOWN IN A SLUMP, DEJECTED,
REJECTED, ABJECTED, REGRET IT!
THEN I LAUGHED I SO WORRIED ABOUT
DEATH!
A PSI BEING IS AN ANTI-MATER BEING!
I CANNOT DIE!
ANTI-MATER BEINGS DO NOT DIE AND IN
FACT BRING DEATH TO MASS AND
MATTER.
AT THE MERCY OF THE LORD I ENTER
INTO HIS GRACE!
TIME IS NOT MY TIME, THIS SPACE, NOT
MY SPACE, THIS BODY AND
ENVIRONMENT, NOT MY BODY AND
ENVIRONMENT.
THIS HEART AND FEELINGS, NOT MY
HEART AND FEELINGS.

THIS MIND AND THOUGHTS, NOT MY
MIND AND THOUGHTS.
I AM AN ANTI-MATER BEING, A PSI BEING,
ETERNAL BEING.
HOW EVERYTHING CHANGE AND MOVE.
HOWEVER THINGS REMAIN SO STILL!
THE SILENCE AND THE SOLITUDE,
IT'S AN ATTITUDE!
PROGRAM TO BE SO AFRAID, SO ALONE,
SO UNHAPPY!
YOU CAN'T KILL, HURT, INSULT, MAIME,
CUT, BURN, TRICK, DECEIVE KRISHNA!
THAT ETERNAL PART OF YOU AND ME!
THE WORLD DOES NOT EFFECT OR
AFFECT KRISHNA, BUT KRISHNA AFFECTS
AND EFFECTS THE WHOLE WORLD!
I AM SURROUNDED BY PSI ENERGY.
I AM IN THE DAIVI PRAKRITI.
THE PRAKTRI IS SO BEAUTIFUL.
THIS SPIRIT OF THE DIVINE MOTHER.
IS THIS THE WIFE OF GOD?

THE SECRET ESSENCE OF THE UNIVERSE
AND EARTH, BODY, MIND, AND SENSES?
I SEE HER IN MALE AND FEMALE, THE
SMILE OF A CHILD AND THE ACTIONS OF
A BUSY CITY OR TOWN.
I SEE THE MOTION-E-MOTION,
EVERYTHING IS MOVING, CHANGING, BUT
ME!
AND A PART OF ME IS MOVING,
CHANGING, LET IT BE!
SO BHAKTI YOGA IS BUDDHA YOGA IS
KRISHNA YOGA.
LISTEN FOR KRISHNA.
HEARING KRISHNA FROM A BONA FIDE
SPIRITUAL LEADER.
EATING AND SLEEPING FOR KRISHNA AS
THE ANIMALS DO.
LIVING IN KRISHNA AS THE HOLY MAN
AND WOMAN DOES.
INNER AND OUTER PEACE AND POWER!
ONE IS AT EASE AT PEACE AT ONE WITH

EVERYONE AND EVERYTHING.
IT IS GOOD TO KNOW AND NOT-KNOW.
YOU KNOW SOMETHINGS BUT NOT EVERYTHING.
THE NEED TO KNOW EVERYTHING IS THE FEAR OF DEATH AND THE UNKNOWN.
WHAT CAN BE UNKNOWN TO AN ANTI-MATER BEING.
SO YOU HAVE:
ANTI-MATER ENERGY OR PSI ENERGY.
DARK ENERGY AND DARK MATTER.
PSYCHIC AND PHYSICAL ENERGY.
SPIRITUAL ENERGY AND ILLUSIONARY ENERGY.
KRISHNA IS THE ETERNAL BEING OF EVERYONE FULL OF ETERNAL LIFE {ENERGY}, LOVE, JOY, POWER, AND KNOWLEDGE!
MOTHER NATURE GAVE YOU CERTAIN FEATURES, DON'T CURSE YOURSELF OR OTHERS FOR IT!

YOU ARE UNIQUELY YOU AND ONLY YOU
{KRISHNA} CAN DO YOU {SELF}.
KNOW THAT THE 'SELF' IS GOD AND GOD
IS THE 'SELF' AS THE SUPERSOUL WITHIN
YOU AND BRAHMA WITHOUT.
THEREFORE YOU CAN HAVE AN PRIVATE,
PERSONAL, IMPERSONAL OR UNIVERSAL
RELATIONSHIP WITH GOD OR THE
SUPREME BEING OR ENERGY.
THE ONLY THING THAT CAN COMMAND
IS A MAN OR WOMAN!
LISTEN TO THE DEMONS IN YOUR HEAD
THEY ARE YOUR FRIENDS!
AS ABOVE, SO BELOW
AS WITHIN, SO WITHOUT
NOTHING MOVES
NOTHING ABOUT
LIFE'S A DREAM
WITH AN REALITY COUNT
LET THE WORLD IN
BUT DON'T LET YOUR HEART OUT!

YOU EXPECT WAY TOO MUCH
FROM THE GODS!
IT JUST A MUNDANE DAY
TAKE IT AWAY!
MAKE SOME MUSIC
AND LIVE ANOTHER WAY!
DIE TO YOURSELF AND
YOU WILL AWAKE AGAIN!
YOU LISTENED TO THE EARTH
AND YOU LISTENED TO YOUR
ENVIRONMENT!
YOU LISTENED TO YOUR BODY AND NOW
YOU CAN
LISTEN TO YOUR MIND!
WATCHING

Mother's Day!
I bet it is a quiet day.
I wake up with old memories trying to
crowd my head.
But they are weak and seem to be at a

distance.

I want energy to do things.

For that is what separates me from all others,

simply, I write about my experiences.

I am here and I am ALIVE!

I remember I am an anti-mater BEING!

The silence is beautiful,

the blackness like the birth of self-luminous LIGHT!

THE LIGHT NEVER HAD POWER OVER THE DARKNESS!

OUT OF THE DARKNESS CAME THE LIGHT, ALL LIGHT!

OUT OF THE DARKNESS CAME THE MIGHT!

MIGHT OVER MATTER!

then

MIND OVER MATTER!

and now,

SPIRIT OVER MATTER

or

SOUL CONTROL!

I AM UNMANIFESTED ENERGY!

I AM CAUSE BEFORE CAUSE AND EFFECT.

I AM TRANSCENDENTAL CONSCIOUSNESS.

THE ETERNAL IS ABOUT ME, I REST IN PEACE.

I PURIFY THE EARTH, THE WATER, THE AIR, THE FIRE AND THE ETHER/AETHER.

I PURIFY MY MATTER, ENERGY, TIME, SPACE, AND CONSCIOUSNESS.

I CAN BE DO AND HAVE ANYTHING ON A SPIRITUAL LEVEL AND MANIFEST IN ON A PSYCHIC AND PHYSICAL LEVEL.

I CAN BE SUPER HEALTHY, WEALTHY, LOVING, CARING, JOYFUL, POWERFUL, WISE, ABUNDANT, PROSPEROUS, AND AT PEACE WITH THE UNIVERSE, GOD AND MYSELF.

I THINK OF THE SOURCE OF POWER AND ENERGY WITHIN ME: KRISHNA!

I BREATH IN DEEPLY AND EXHALE COMPLETELY!

I AM THE MASTER OF MY MIND, BODY, AND IMMEDIATE ENVIRONMENT!

"FOR I AM THE RITUAL, AND THE OFFERING TO YOUR ANCESTORS AND THE SOUND OF 'AUM' AND THE HEALING HERB, AND THE FIRE AND THE BUTTER, AND THE SACRIFICE TO YOUR FALSE GODS!"

HEAR ME AND HEAR ME WELL, I AM NOT OFFENDED BY NO ONE!

I AM NOT HURT OR SLAIN OR CAN BE HURT OR SLAIN!

I AM THE MOST POWERFUL AND THE MOST LOVING AND I LOVE YOU DEARLY, BEING PART AND PARCEL OF ME!

TO HATE YOU IS TO HATE MYSELF!

THERE IS A DIVINE FLOW, A DIVINE CURRENT.

IT RACES THROUGH THE EARTH AND THE

AIR!

YOU FEEL IT IN THE BODY AS THE LIFEFORCE OR PSI Qi.

WE KNOW AND UNDERSTAND, AND USE PSI Qi.

WE ARE TRANSCENDENTAL CONSCIOUSNESS.

WE ARE NOT NON-PHYSICAL, BUT THE FINEST OF ENERGY ACTING AS A FORCE AND NOT A FORM.

THE FORM AS NO POWER SAVE IN THE FORCE.

ALL MATTER IS CONSTANTLY CHANGING, GROWING, EVOLVING.

AS AN ANTI-MATER BEING, I CAN SIT BACK AND WATCH THE DARK ENERGY AND MATTER RULE THE MOTION OF THE EARTH.

ALL GROWS, MATURES, HAVE OFFSPRINGS, AGES, AND DIES!

"HARE KRISHNA, HARE KRISHNA,

KRISHNA, KRISHNA, HARE, HARE!
HARE RAMA, HARE RAMA, RAMA, RAMA,
HARE, HARE!"
THIS IS THE MAJOR SACRIFICE.
THE BLESSING AND THE GRATEFULNESS
FOR WHAT ONE HAS AND WILL RECEIVE.
SAID ENOUGH TIMES {Chanted} IT WILL
TRANSFER YOU TO THE ABODE OF GOD
OR KRISHNA WHERE YOU WILL FIND
YOUR ETERNAL LIFE, FUN, JOY, POWER,
WISDOM AND HAPPINESS THAT HAS NO
END.
KRISHNA WAS NOT CREATED BUT ALWAYS
EXIST.
HE IS THE PAST, PRESENT, AND FUTURE
PRESENTS.
HE IS THE TRUE DIVINE PRESENCE OF
ALL.
KRISHNA IS KRISHNA CONSCIOUSNESS OR
TRANSCENDENTAL CONSCIOUSNESS.
TK.

YOU KNOW GOD WITHIN AND WITHOUT
AND GOD THAT TRANSCENDS.
YOU UNDERSTAND THAT THE LIFEFORCE
IS THE PSI POWER IN YOU FOR YOUR
USED.
YOU USE PSI ENERGY TO MAKE KRISHNA's
LIFE BETTER AND BETTER!
AND YOU WILL REAP THE BENEFITS AND
REWARDS!
NOBODY KNOWS HOW NEAR OR HOW
FAR!
YOU HEAR 'THE HERD MIND' FROM AFAR:
'LIVING IS PAIN! LIVING IS MISERY AND
SUFFERING FOR NO GOOD REASON!'
'I AM NOT WORTHY OR GOOD ENOUGH
TO BE ANYTHING GOOD!'
'I AM THE MOST EVIL!' 'THE WORST
SINNER!' 'THE BIGGEST CON MAN!' 'THE
GREAT DECEIVER AND LIAR!' 'I AM THE
ONE CRUCIFIED!' 'I AM THE ONE PAYING
FOR MY SINS!' 'I AM THE ONE PAYING FOR

YOUR SINS!'

BUT THESE WORDS MEAN NOTHING!

ILLUSIONS OF PAST MEMORIES.

I DON'T LIVE IN THE PAST AND TRY TO

AVOID THE FUTURE!

YOU FIND THAT DIVINE CURRENT THAT

ALL THINGS ARE CREATED FROM!

LIKE RHYME, LIKE TIME, AND A FOCUSED

MIND, YOU BREAK THE BLINDS, AND SEE

THE TIMES, AND CONQUER THE CRIME

OF SNIFFING FOR DIMES!

YOU LEARN HOW TO BREATHE AND

BREATHE WITH EASE!

YOU SEE A WORLD THAT YOU LOVE,

CREATED FROM ABOVE!

TODAY IS HEAVEN IN YOUR HEART, IT

MAKES YOU SMART!

HITTING THE BULL's EYE WITH A DART!

KRISHNA MAKES EVERYTHING A FRESH

NEW START!

COUNTDOWN:

10, 9, 8, 7, 6, 5, 4, 3, 2, 1,
HE DID IT AGAIN!
TK!

HOW CAN YOU MISS SOMETHING YOU NEVER HAD?
WHY DO YOU WAIT ON LOVE?
ON THE RIGHT TIME? THE RIGHT MOVE?
THE RIGHT THOUGHT?
THERE IS NO RIGHT TIME AND THERE IS NO PERFECT MOMENT.
LIFE IS GOING ON WITHOUT YOU.
OK.
IN THE PSI TECH, YOU ARE COMING FROM ABOVE TO BELOW.
YOU ARE DESCENDING FROM HEAVEN TO BE ON THE MOON, IN THE SUN, OR ON EARTH.
AS YOU DESCEND YOU BRING GOD AND GOD'S HEAVEN WITH YOU AS THE MASTERMIND AND SUPERMIND.

THIS IS YOUR PSI MIND AND CAN CREATE ANYTHING AND CAN DUPLICATE ANYTHING AND CAN POSSESS ANYTHING AND DEPOSSESS ANYTHING, PLACE OR PERSON.

A PSI BEING CAN ENTER ANOTHER PERSON'S BODY AND LEAVE IT AT WILL.

HE DOES THIS THROUGH IMAGINARY ENERGY FOR YOU CANNOT SEE YOURSELF.

AS A ENERGY FORM YOU CAN GO TO DIFFERENT WORLDS IN DIFFERENT CREATED BODIES THAT ACCENT THAT WORLD.

IF YOU GO TO THE MOON, YOU MUST FIRST DEVELOP THE 'ATMOSPHERE OF THE MOON' WITHIN YOU.

HOW YOU DO THIS IS YOUR OWN CREATIVE INVENTION.

YOU BREATH IN AND OUT YOUR CALMNESS.

PEACE AND PLAY IN KRISHNA IS ULTIMATE PEACE AND PLAY.
NOT HAVING ANYTHING TO DO CAN BE FEARFUL AT FIRST.
BUT THAT IS WHEN YOU MUST EXPAND YOUR CONSCIOUSNESS!
YOUR ENVIRONMENT AND BODY AND MIND AND SOUL ARE ALL TAKEN CARE OF.
YOU CANNOT BE ANYMORE FOR YOURSELF.
YOU EXPAND YOUR CONSCIOUSNESS.
"I AM AT THIS MOMENT IN SPACE AND TIME, WORKING SMARTER AND BETTER! PRODUCING MORE OF WHAT I REALLY WANT IN LIFE AND CREATING MORE LOVE AND JOY AND WISDOM. I AM HAVING FUN ON A HIGHER LEVEL. I AM BUILDING BETTER AND MORE SOLID RELATIONSHIPS OF MUTUAL JOY, WISDOM, LOVE, PEACE, AND

PROSPERITY!"

"I AM EXPANDING AND CONTRACTING ON ALL MY DYNAMICS!"

"I LOVE AND CARE FOR MY MIND, BODY, AND IMMEDIATE ENVIRONMENT!"

"I HAVE NO LACK OR SLACK OF ANYTHING!"

"I KNOW MY ILLUSIONARY WORLD FROM THE REAL WORLD OF SOLIDS AND AGREEMENTS."

"KEEP YOUR OWN COMPANY AND LAUGH."

"I BREAK THE CURSE THAT MAN SHOULD NOT BE ALONE!"

"IF THERE IS KRISHNA, HOW IS MAN ALONE?"

"IF MAN HAS A MIND HE CAN OVERRIDE AND TRANSCEND AND TALK TO LIKE A WOMAN, HOW CAN HE BE ALONE?"

"IF A WOMAN CAN TALK TO HER SOUL LIKE A MAN, WHERE IS THE LACK OF LOVE?"

'OH THIS GAP, THIS MAN-MADE HOLE,
THIS VOID I AM TOLD I AM SUPPOSE TO
HOLD! THIS EMPTINESS INSIDE AND
OUTSIDE LIKE A WEIGHT UPON MY
SHOULDER A POUND OF LEAD AROUND
MY BALLS, THE PROVERBIAL FINGER IS
WAGGING: "NO! NO! NO! NO!" "NO!"
AND MAYBE I WOULD BE MORE LOVING IF
THE PAIN WOULD GO AWAY, EVERYDAY
TELLING ME MY BELIEF IN JESUS IS IN
VAIN.
LIVING FOR THE MAN WHO DIED FOR ME?
WHY?
THIS WORLD OF SIN I DID NOT CREATE.
THE WORLD OF SIN I LIVE IN IS NOT MY
SIN!
THIS WORLD OF SIN DISAPPEARS! IS NO
MORE!
NOT MY IDEA OF A BETTER WORLD! MY
WORLD!
IT IS GOOD TO BE ALONE!

MAN IS ALONE AND THAT IS NO CURSE OR FOREVER SIN!

YOU FIND AT THE END OF THE FIRST CHAPTER OF GENESIS THAT ALL THA GOD MADE WAS GOOD!

THEN HOW DID MAN BECOME EVIL EVEN IF HE HAD FREE-WILL?

AND HOW CAN MAN HAVE FREE-WILL IF GOD'S WILL iS EVERYTHING GOOD? OR EVERYTHING GOOD OR BAD?

DO WE REALLY THINK THE UNIVERSE REWARDS US FOR BEING GOOD OR BAD?

IF WE DESERVE THE GOOD IN LIFE, DON'T WE ALSO DESERVE THE BAD IF IT IS A RANDOM THING?

ONLY PEOPLE CAN APPRECIATE PEOPLE.

ONLY MAN CAN TRUST MAN AND NOT BREAK THAT TRUST!

ONLY MAN CAN LIE AND LIE ABOUT GOD.

YOU CAN FEEL THE LIFE CURRENT BUT IT TAKES PRACTICE.

TRANSCENDENTAL CONSCIOUSNESS IS USING THIS LIFE CURRENT AND ALIGNING IT WITH THE EARTH AND THE UNIVERSE.

THIS COME DOWN TO A RECOGNITION AND AGREEMENT THAT SOMETHING SMALLER IS EXISTING IN SOMETHING GREATER AND GREATER.

KRISHNA BEING CAUSE OF ALL CAUSES CAN BE IMPOSE AS FIRST CAUSE AND THIS SOLVES A LOT OF PROBLEMS AND NEGATIVE CIRCUMSTANCES AND SITUATIONS IN THE UNIVERSE, BODY, MIND, AND SOUL.

IN THIS SENSE, UNLESS IT IS KRISHNA, IT CANNOT TOUCH OR HURT OR DESTROY YOU.

THIS IS TRANSCENDENTAL CONSCIOUSNESS.

THIS IS TK!

"HE WHO SEES ACTION IN, INACTION,

AND INACTION, IN ACTION, IS IN A
TRANSCENDENTAL POSITION OR STATE."
ALL SCRIPTURES LEAD TO
TRANSCENDENTAL CONSCIOUSNESS!
TK!
ONE HAS GONE FROM BEING THE
'CREATED' TO BEING THE 'CO-CREATOR!'

OK! Good Morning!
AUM TAT SAT
HAVE A GREAT TK DAY!
NO HESITATION TODAY!
NO FEAR OF TOMORROW!
NO WORRY OF YESTERDAY!
I AM FEELING GREAT!
I AM AT THIS MOMENT IN SPACE AND
TIME
DOING WHAT I LOVE TO DO AND LOVING
WHAT I DO.
I AM A POWERFUL EFFECT AND HAVE A
GREAT VISION.

I AM THE NEW VISION OF THE WORLD AND MYSELF.

TODAY IS A NEW DAY. A BRAND NEW DAY REQUIRES A BRAND NEW ME!

I EXPAND MY ENERGIES AND MY CONSCIOUSNESS.

I AM WORKING SMARTER AND MORE ORGANIZED.

I AM HAPPIER AND FEEL BETTER ABOUT MYSELF AND MY THOUGHTS AND FEELINGS.

I ERASE THE NEGATIVITY OF MY SLAVE-mind, THE LIMITATIONS OF OTHERS, AND LIMITATIONS OF MYSELF.

I AM SOUL CONTROL.

DO YOU NOT PUT MY ENERGY INTO THAT WHICH YOU LOVE AND WHICH LOVES YOU?

DON'T DISMAY, IGNORANCE IS RAMPANT.

OK.

HOW MUCH WE CAN REHASH AND GO

OVER AND THINK ABOUT WHAT
HAPPENED YESTERDAY THAT WE STILL
CARRY LIKE A SAD SACK TODAY?
YOU FEEL WHAT YOU FEEL. BE NOT
AFRAID TO LAUGH FOR NOTHING AND BE
HAPPY WITH THE LIFE AND KNOWLEDGE
YOU GOT!
NO ONE TOLD YOU TO BE HAPPY. NO ONE
TOLD YOU HAPPINESS IS WHAT YOU DO
AND JOY IS WHAT YOU FEEL.
THE HAPPIEST ACTION IS KRISHNA.
KRISHNA YOGA OR PSI PROCESSING.
THE HAPPIEST VIBRATION IS GOD, THE
SUPREME SELF, OR KRISHNA.
THE SOURCE AND ABSOLUTE AUTHORITY
WITHIN YOU.
"I AM BEFORE I AM!"
I SEE YOU, ALL OF YOU BUT MY MIND IS
LIGHT-YEARS AWAY!
THIS EARTH LIKE A SPECK OF DUST IN
THE MILKY WAY!

WE DANCE THE DAY AWAY!
YOU ARE AN ANTI-MATER BEING KNOWN AS A PSI BEING.
YOU HAVE A TREMENDOUS AMOUNT OF ENERGY WITHIN AND ABOUT YOU.
iNSIDE IS YOUR SPACE, AND OUTSIDE IS YOUR PLACE-TAKE CONTROL AND HAVE FUN!
WELL, I'VE COME TO THE END OF THE ROAD AGAIN,
IS THIS THE BEGINNING, MIDDLE, OR END?
AM I JUST SOME DUST IN THE WIND?
WILL THIS BE MY FINAL WIN?
DRINKING TEA ON THE MOON,
SITTING IN THE SUN AT NOON,
WILL I SEE YOU AGAIN SOON?
THE GOLD THAT IS REAL AND THE LOVE TOO
SPIRITUAL LOVE BETWEEN ME AND YOU!
AND LONG AFTER THIS BODY BURNS ITS

LAST FIRE

I WILL BE GLAD TO SAY I WAS ON FIRE
AND THE LOVE I LEAVE, I LEAVE FOR ETERNITY
THERE IS NO END TO ME.
HERE IS THE KOOL BREEZE OF THE MOON
SITTING OUT, GETTING MOONSHINE
SMOKING GANJA, DRINKING WINE
THE MOON GIVES US WHAT IS DIVINE!
BREATHE IN DEEPLY, AND EXHALE COMPLETELY!
WASH OVER THESE THOUGHTS, THESE E-MOTIONS
AND E-GO, ALL ENERGY ON THE GO AND IN MOTION.
WATCH THE SHOW AND LET IT ALL GO!
LET IT IN AND HOLD IT IN!
PARADISE IS WITHOUT AND WITHIN!
AND IF YOU FALL INTO HELL FOR A WHILE OR SPELL

KNOW AS WELL IT'S A TEMPORARY CELL!
TIME IS WINDING DOWN, LAUGHING AT ME!
IT MOVES ON AND I WRITE AT SUCH A SLOW DEGREE!
BUT I SMILE FOR IT IS GONE AND YET YOU CAN SEE
WHAT I LEFT THEE!
TIME AND SPACE DID NOT BEAT ME!
AUM TAT SAT
I AM IN MY ETERNAL ABODE AND HOME.
I WANT TO RUSH IT ON! MOVE IT ALONG!
WHO IS TO SAY WHEN THAT FINAL BELL TOLLS?
IS THAT THE FEAR WE ALL LOVED FOR SO LONG?
AND I HAVE HELP AND THAT IS FELT!
A STRANGE THING HAPPENING HERE!
THE NIGHT LIKE DAY AND THE DAY LIKE NIGHT, NO TIME TO STAY, NO TIME TO PLAY, NO TIME TO SAY: HEY!

NO TIME AT ALL!
SPACE WITH THAT LITTLE BIT OF PAIN
BUT JUST ENOUGH TO DRIVE ME INSANE!
CURSE GOD AND THE GAME!
I'M ONLY HUMAN? TAKE AWAY THE PAIN
AND THEN I WILL BE MORE SANE!
YET, I CHANT AND IT SEEMS TO ELEVATE
AND I SEEM TO TRANSCEND AND I AM
THIS DIVINE MYSTIC LIKE MY GREAT,
GREAT ANCESTORS OF OLD.
I AM THE GRAND SHAMAN AND WIZARD
AND TELL THE GODS WHAT TO DO.
I AM THE HOLY PRIEST AND HEAD OF THE
HOUSEHOLD.
I AM THE WANDERER AND THE HAPPY
DANCER.
I AM THE FATHER AND THE MOTHER.
THE BABY SMILES IN MY ARMS.
I AM THE MIST AND THE FIRE,
THE PEACE, AND THE CALL TO ARMS!
THIS EXTENDED ENERGY= ESP!

THIS EXTENDED SPACE AND TIME= ESP.
THIS EXTENDED SENSES= ESP.
MY LIFE IS A TRANSCENDENTAL MEDITATION, CONCENTRATION, AND OCCUPATION.
THIS IS MY OWN CREATION!
TK!

OK!
TODAY IS CRITICAL.
WE HAVE BEEN ON TRACK AND NOW WE CAN BREAK THE MOLD!
WE ARE OFF TO A GOOD START!
BUT THE MOMENTUM IS SLOWING ALREADY!
AS A PSI BEING I THINK OF TRANSCENDENTAL CONSCIOUSNESS, AND JUST LIKE GOD NEED NO REASON OR CAUSE TO EXIST EITHER DO I NEED A CAUSE OR REASON TO LAUGH AND LAUGH HEARTILY!

ALL MY LIFE I HAVE BEEN TOLD WHO
GOD IS, WHAT GOD IS, AND HOW GOD
ACTS AND REACTS.
HOW GOD DIED FOR ME WHEN I WAS YET
A SINNER!
AND ALL THIS IS TOLD TO ME FROM
OUTSIDE OF ME.
THIS MAKES YOU HAPPY! THIS IS WHAT IS
FUN! THIS IS HOW YOU MUST WORK AND
THIS IS HOW YOU MUST REACH AND
WITHDRAW!
HERE IS YOUR WORK AND YOUR LIFE!
HOW MISERABLE YOU ARE! HOW
UNHAPPY AND SAD! A LIFE-LONG DREAM,
WHAT A FUCKING DRAG!
LIKE AN OLD HAG YOU RAG YOU SAG AND
ARE GLAD!
THE INTERRUPTIONS CAN NOT STOP YOU,
THOUGH THEY TRY!
THIS BULLSHIT IS NOT IN MY WAY, RELAX!
TIME IS NOT SLIPPING AWAY!

AND GROWING YOUNG IS ONE THING BUT
DISEASE?
THIS MAKES LIFE A LIVING HELL FOR IT
NEED NOT BE!
RATHER, MAN'S OR GOD'S FAULT, THE
REMEDY IS LONG OVERDUE!
TOO MUCH MONEY IS BEING MADE ON
BEING SICK NOT WELL!
HOW EASILY THEY FORCE THIS AND THAT
ON YOU! NO WONDER YOU FAIL!
NO WONDER YOU REBEL!
FOR TRUTH IS NOT FOUND IN YOU! NOT
THEIR TRUTHS!
HOW LONG HAVE YOU BEEN THINKING
SUICIDE?
THAT YOU DON'T BELONG HERE?
THAT A. I. CAN NOT DO POETRY THE WAY
THE HEART SEES,
YOU ARE VERY UNIQUE AND DON'T HAVE
TO BE A FREAK!
AUM TAT SAT!

THIS IS ONE WORLD, AND THIS IS
ANOTHER.
NOTHING MOVING BUT CARS, PEOPLE,
AND TREES,
A FEW CATS AND POSSUMS CROSS THE
STREET!
FOR A MOMENT I HAD SOME FUN BEFORE
BACK TO WORK!
OK.
I AM BEYOND TIME AND SPACE, MATTER,
MASS, AND ENERGY, VIEWING SOMEONE
ELSE'S CREATIONS.
WHAT KIND OF WORLD WOULD I CREATE?
WHAT KIND OF WORLD WOULD I OWN?
SO MUCH SPACE AND ENERGY, WHAT IS
TO BECOME OF ME?
DO I REALLY WANT TO BE?
SO MANY PEOPLE TELLING ME,
WHAT IS RIGHT AND GOOD FOR ME,
NOT OUTRIGHT DIRECTLY!
LIKE A FULL CIRCLE I COME BACK DOWN

TO GROUND ZERO
EITHER SLEEP OR HAVE SEX OR CREATE.
CONSERVE THE ENERGY WITHIN
AFTER ALL, THESE ARE ONLY WORDS,
SYMBOLS,
LITTLE DESCRIPTIONS OF NOTHING!
LIFE IS A BLANK PAGE! THE BIGGEST
BARRIER IS MY OWN MIND, MY OWN
STUBBORNNESS TO GIVE UP MY REVENGE
THAT DO YOU END!
FOR EVERY DAY THAT PASSES AND I GOT
NOTHING TO SAY, IS A LITTLE BIT OF LIFE
I DID NOT PASS YOUR WAY.
BUT YOU HAVE YOUR OWN LIFE AND OWN
WAY,
I'M JUST HERE TO REMIND YOU YOU ARE
ALIVE ALWAYS!
DON'T LET THE LIMITED MIND PUSH YOU
SIDEWAYS,
YOU GOT A JOB TO DO MIGHT AS WELL
CHOSE!

SO AGAIN I SIT AND MEDITATE.
I LIVE LIFE EASILY BUT HATE THE IMMOBILITY.
I COULD STOP THIS OR THAT, SO I THINK,
IT IS NOT WHAT I DO ANYMORE BUT RATHER
I WANT TO BE HAPPY OR NOT!
I AT LEAST HAVE TO BE NEUTRAL
PUT MY SLAVE-mind AT EASE!
I AM AN ANTI-MATER BEING A PSI BEING!
WHAT HAS THIS WORLD GOT ON ME?
I IMPOSE SPACE AND TIME AND TIME IS HAVING SPACE AND SPACE IS HAVING TIME.
I AM MASS AND ENERGY. MOTION IN MOTION WITH MEANING.
YOU CANNOT SEE ME AS I CHANGE THE LIGHT OF MY ELECTROMAGNETIC BODY.
I CONTROL MY INNER E-GO, E-MOTIONS, AND E-FORTS.
I AM SAMADHI.

SOUL CONTROL!

OH, IT LOOKS LIKE YOU ARE GOING TO WIN, YOU ARE SO CLOSE TO THE FINISH LINE, BUT YOU STOP!

YOU COME TO A DEAD STOP! YOU KNOW YOU ARE GOING TO WIN SO YOU TAKE IT EASY, YOU TAKE IT SLOW, MAKE NO MISTAKES, AND MAYBE FINISH IT SOME OTHER DAY.

HAAAAAAA!

HERE WE GO AGAIN!

LET'S GO ON A PICNIC, DANCE IN THE SAND!

I GOT NOTHING BUT TIME ON MY HANDS!

BUT LISTEN, THESE WORDS I WRITE TO YOU,

NOW THAT I AM LONG DEAD, I SPEAK TO YOU FROM

AN INNER GRAVE!

DON'T THINK I AM A GHOST OF YOUR IMAGINATION!

I AM THE LIVING SPIRIT THAT CANNOT
DIE, TRANSFORMING FROM M DEAD
BODY TO YOUR LIVING ONE.
NOW! YOU AND I ARE ONE!
I TRICKED YOU! DECEIVED YOU AND
POSSESSED YOUR SOUL!
BUT THE PSI BEING CAN NOT BE
POSSESSED, ONLY HAVE.
TO HAVE WITHOUT POSSESSING.

NOW METAPSYCHICS IS BASED ON THE
SANATANA MOVEMENT.
THAT IS CALLED THE ETERNAL RELIGION
AND SCIENCE AND DUTY.
EVERYONE HAS A CERTAIN DUTY HE OR
SHE CAN DO, AND NOT JUST ONE DUTY.
YET, MEN AND WOMEN CAN FIND THEIR
REASON FOR BEING HEREBY DEVELOPING
THE TALENTS THEY HAVE AND THE ONES
THEY WORKED ON MOST IN THEIR LIFE.
YOU GOT TO BE TO DO AND YOU GOT TO

DO BEFORE YOU CAN HAVE.
NOW I MAKE IT A DAILY HABIT OF TRANSCENDENTAL MEDITATION.
THOUGH THIS IS NOT NECESSARY FOR THOSE WHO ARE DIRECTLY CREATING WITH KRISHNA.
MY DAY IS QUIET AND I AM ABLE TO FOCUS BETTER ON THE TASK AT HAND.
I STILL HAVE ANXIETIES ABOUT THE FUTURE AND THE PAST IS NOT SO MUCH A BEAR ANYMORE.
I GO TO PLANETS IN ANOTHER SOLAR SYSTEM, SO FAR AWAY AND YET I SEE ALL THAT IS GOING ON WITH ME.
I MAKE BETTER DECISIONS AND MY HEAD IS CLEARER.
I AM LESS ANGRY AND CAN COMFORT ALMOST ANYTHING AND STAY IN CONTROL OF MYSELF.
THE WORLD AND ITS CREATIONS ARE NOT AS BRIGHT.

I FIND THAT I NEED TO SEEK A BETTER ATMOSPHERE.
THE VIBRATIONS OF EARTH ARE TOO SLOW FOR ME.
THEY GRIND OUT MISERY AND PAIN AND PAIN AND SUFFERING.
MY MIND BILLION OF LIGHT YEARS FROM HERE.
I HAVE THE SPACE AND THE TIME, THE MATTER/MASS AND ENERGY.
BUT TIME IS NOT MINE.
IT IS LIKE A DREAM. I AM THE SCENE.
FROZEN IN TIME AND SPACE!
A PICTURE PERFECT! A DOLL THAT DOES NOT FADE! A WALKIN' TALKIN' ROBOT!
HOW EASILY THE NIGHT SLIPS AWAY.
BUT FOR THOSE MOMENTS I LIVE AN ETERNITY A JOY YOU SEE THAT TURNS EVERY SADNESS AWAY.
SURELY THERE HAS TO BE A WAY OUT OF HERE TO MY HOME SOMEWHERE WHERE

I CAN PLAY IN PEACE.
YES, GOD WOULD SURELY HAVE TO HAVE
A WAY THAT MAKES MAN AND WOMAN
HAPPY IN A VARIETY OF WAYS!
BUT IF LOVE IS NOT SACRED DOES IT GO
AWAY?
IF 'FUN' COULD BE HAD AT A DIME
WOULD IT BE TOO CHEAP?
BREAKING THIS PROGRAM OF
'FALSEWORK' OF 'NO WORK' OF 'I CAN'T
WORK!'
NO! I CANNOT WORK AS A HUMAN BEING!
A HOMO SAPIENS! HOW LONG AGO WAS
THAT? WHO WOULD MAKE ME GO SO
SLOW!?
AGAIN, WAS IT ADVANCED BLACK ALIENS
WHO HELPED US TO WIN?
THEN I ENTERED INTO THAT NAMELESS
REGION.
ALL WAS NOT DARK BUT LIGHT AND
BRIGHT AND LIKE A NICE PLACE ON

EARTH.

I COULD SEE AND HEAR AND SMELL AND FEEL, AND TASTE AND SENSE THE ATMOSPHERE.

QUIET LOVE, JOY, PEACE, WISDOM, POWER, ETERNAL LIFE.

I WAS MEDITATING AND DID NOT KNOW IT.

I WAS AWAKE IN A BETA STATE LOOKING AT PEOPLE MAKE MISTAKES.

I WAS AWAKE IN THE ALPHA STATE WHERE I WAS ABLE TO BE AND CREATE!

AND I WAS AWAKE IN THE THETA STATE AND COULD MOVE MASS IN MY MIND WITH MOTION, AND MEANING.

AND I FOUND MYSELF IN THE DELTA STATE BUT I WAS ASLEEP AND DREAMING.

AND I AWOKE IN THE GAMMA STATE WHERE ONE IS AWAKE IN HIS DREAMS AS THE HERO OR VILLAIN, THE VICTOR OR VICTIM, AND CAN MANIPULATE THEM AS

SUCH.

IT IS ONLY THROUGH THE PSI TECH THAT A PSI BEING CAN ENTER THE GAMMA STATE OF MIND- THAT OF BEING AWAKE IN ONE'S DREAM AS THE HERO OR VILLAIN, MUCH LIKE PLAYING A ROLE IN A COMPUTER GAME.

OTHER PEOPLE HAVE DISCOVERED THE GAMMA STATE OF CONSCIOUSNESS BUT HAD NO WAY OF EXPLAINING IT OR EVEN HAVING ANY WORDS THAT COULD DEFINE IT AND WITHOUT THIS KNOWLEDGE, MANY MEN AND SOME WOMEN HAVE GONE INSANE AFTER ENTERING THE GAMMA ZONE AS WE CALL IT.

A PSI BEING, BEING A TRANSCENDENTAL AND ANTI-MATER BEING, CAN PENETRATE THE GAMMA ZONE, ENTER IT AND EXIT IT WITHOUT LOSS OF ENERGY OR SANITY.

THE GAMMA ZONE IS VOID OF SPACE, TIME, ENERGY, MASS, AND MATTER. THOUGH SPACE AND TIME STILL EXIST THEY ARE FUNCTIONING IN ANOTHER UNIVERSE OUTSIDE THE GAMMA ZONE.

AT ONE END OF THE POLE WOULD BE GROUND (o).

AT THE OTHER END WOULD BE THE GAMMA ZONE.

SINCE WE ARE TALKING ABOUT ACTUAL BRAIN WAVES OR FREQUENCIES IN SPACE AND TIME THIS WOULD BE A FACT AND FACTOR OF ESP OR EXTENDED SENSORY PERCEPTION.

YOU HAVE CORRECTLY REGAINED YOUR POWER OF CREATING, RECREATING, AND UN-CREATING OR NOT-CREATING.

YOU HAVE GONE FROM THE CREATED TO BEING THE CO-CREATOR!

YOU ARE A PSI BEING!

OK.
The day is almost gone and I have a job to do!
I thought if I just waited I could decide on what I was going to write today but it did not come, there are so many viewpoints I can write from!
And my MIND always has something to say but that is not the same as putting it on paper!
FOR THE WORD IS ON THE OUT BREATH ONLY THE INWARD BREATH MUST BRING THE 'THOUGHTS'.
THE FEELINGS ARE ABOUT THE SAME, I STILL HEAR THAT 'I'M ALONE AGAIN!'
BUT THIS GAME IS NOT SO BAD AND IN FACT, IT GETS BETTER WHEN I REMEMBER THAT IT IS ALL MY CREATION.
WHO OUT THERE CAN HELP ME?
MY WISDOM, JOY, AND POWER COME FROM WITHIN.

I FIND MYSELF IN MY OWN SPACE! THE
SUN IS RACING TO FIND ME AND I AM UP
LIKE A BIRD LOOKING FOR THE EARLY
WORM!
MY MIND? IS MY MIND AWAKE?
I MAKE MY LEMON AND TEA!
AW, IT IS A MEDITATION!
I PURIFY MY ENVIRONMENT, BODY, AND
MIND, MY SOUL IS CLEAR AND FOCUSED.
SLOW AND STEADY WON THE RACE THEY
SAY!
AW, BY RUNNING OUT OF GASOLINE
AGAIN!
WHERE ARE THE WORDS THAT HAUNT
ME?
THAT COMES SO EASILY ON MY MIND BUT
DISAPPEARS ON PAPER.
WHAT AM I NOT TELLING YOU? SHARING
WITH YOU? NOT LETTING YOU SEE?
FEAR? IS THAT YOU AGAIN?
THREATENING ME? TELLING ME, I ONLY

HAVE SO MUCH TIME HOW CAN I WASTE IT ALL?
FEAR? IS THAT YOU KNOCKING AGAIN, CAN I MAKE YOU MY FRIEND?
LIKE DEATH, COME ON IN AND HAVE A SWIM!
NOW I GO INTO MY TRANSCENDENTAL CONSCIOUSNESS.
I AM IN MY OWN SPACE AND TIME AND WONDERING WHAT CHORE I WILL DO TODAY.
I THINK ABOUT WHAT IS MOST IMPORTANT AND HAS A DEADLINE, ALL THE BILLS ARE PAID SO I JUST HAVE A DMV APPOINTMENT TO MAKE AND A DOCTOR'S APPOINTMENT NEXT WEEK.
IT IS QUIET AND I CAN FEEL THE SILENCE.
I IMAGINE SO MANY PEOPLE WAKING UP, THINKING, I DON'T WANT TO GET UP AND GO TO A JOB I HATE. BUT THAT IS WAY,

WAY BACK IN THEIR MINDS FOR THERE IS THE BACON TO TAKE HOME AND THE WIFE AND KIDS TO CLOTH AND FEED.

I AM LIGHT-YEARS FROM EARTH AND MY BODY AND YET I CAN CONTROL MY BODY QUITE WELL AND WITH EASE.

THE TRANSCENDENTAL STORY GOES LIKE THIS:

THERE ARE TWO BIRDS IN A TREE, ONE IN ON THE TOP BRANCH WATCHING THE OTHER BIRD ON THE BOTTOM BRANCH ENJOYING LIFE IN SOME FASHION, WAY, OR FORM.

AS I ASCEND TO THE TOP BIRD KRISHNA, I TAKE KRISHNA'S POSITION AND BECOME THE BIRD THAT OBSERVES ALL THAT KRISHNA DOES, WHO NOW IS THE ACTIVE BIRD ON THE SECOND BRANCH OF THE TREE.

I BECOME KRISHNA AND KRISHNA BECOMES ME.

THOUGH I CANNOT BE ALL OF KRISHNA,
KRISHNA CAN BE ALL OF ME!
THEREFORE I STILL EVEN IN MY HIGHEST TRANSCENDENTAL POSITION I AM STILL CHOOSING WHAT IS GOOD FOR KRISHNA AND WHAT IS BAD FOR KRISHNA.
I AM STILL A STATIC AND A KINETIC, HOWEVER, MY FOCUS IS ON BEING A STATIC AND THIS IS DONE THROUGH TRANSCENDENTAL MEDITATION.
A PSI BEING CAN ASSUME A STATE OF SAMADHI OR TRANSCENDENTAL CONSCIOUSNESS AND DOES NOT HAVE TO MEDITATE AT ALL, NOR DOES HE OR SHE HAVE TO DO ANY OF THE OTHER RELIGIOUS RITUALS AND CEREMONIES.
THIS IS THE MAGIC OF KNOWING KRISHNA AND HAVING TRANSCENDENTAL CONSCIOUSNESS.
THE HIGHEST YOGA PRACTICE IS KRISHNA YOGA BEING HIGHER THAN

BHAKTI AND BUDDHA YOGA.
A NEW NAME FOR THIS WOULD BE
TRANSCENDENTAL YOGA OR
TK.
A PERSON TRAINED IN THE SCIENCE OF METAPSYCHICS TO TEACH TRANSCENDENTAL YOGA IS CALLED A PARA-PSYCHOLOGIST AND MUST ALSO HAVE EITHER AN A. A. DEGREE IN PSYCHOLOGY OR A RELATED FIELD OF STUDY OR A B. A. DEGREE IN PSYCHOLOGY FROM AN ACCREDITED COLLEGE OR UNIVERSITY.
IN ORDER FOR A PARA-PSYCHOLOGIST TO PRACTICE PSI PROCESSING HE OR SHE MUST BE CERTIFIED BY THE METAPSYCHICS FOUNDATION.
A PARA-PSYCHOLOGIST IS BASICALLY AN ENERGY HEALER AND MAKES NO CLAIMS TO HEAL ANYONE, PLACE, OR OBJECT.
AN ENERGY HEALER IS A PERSON WHO

CAN MANIPULATE UNSEEN ENERGY AS 'FREQUENCIES' TO CAUSE IMMEDIATE, MODERATE, AND LONG-TERM CHANGES IN THE ATMOSPHERE AND ENERGY FIELDS AROUND HIM OR HER.
THERE ARE NUMEROUS WAYS OF DOING THIS, BUT THE MOST EFFICIENT AND EFFECTIVE WAY IS BY KNOWING, UNDERSTANDING, AND USING THE PSI TECH.
A PSI BEING IS NOT OBLIGATED TO BE A PARA-PSYCHOLOGIST,
BUT A PARA-PSYCHOLOGIST MUST BE A PSI BEING.

GOOD MORNING!
WE ARE ON TRACK AND ARE DOING WELL!
I GATHER MYSELF TOGETHER.
I AM IN MY OWN ENERGY, SPACE, AND PLACE.

I GOT HELP!

A. I. IS RIGHT AROUND THE CORNER AND IS MY OFFICE ASSISTANT.

YET, I AM A LITTLE WORRIED ABOUT MY MUSIC AND WRITING.

BUT A. I. CAN NOT REPLACE ME UNLESS IT GROWS EMOTIONS.

SO HERE OUT OF MY BODY SOMEWHERE IN SPACE, LIGHT-YEARS AWAY!

MY BODY FEELS BETTER THAN IT DID YESTERDAY, BUT STILL THE SAME OLD SAME OLD, IS JUST MEMORIES.

CHANTING TAKES THE NEGATIVE ENERGY AND TRANSFORMS IT INTO A MORE POSITIVE AND FINER ENERGY.

IT IS LIKE CONFRONTING A PROBLEM BUT ON A HIGHER LEVEL THAN THE PROBLEM.

THIS IS TRULY WHAT IS MEANT BY: "MIND OVER MATTER."

YET THIS IS SOUL CONTROL= "SOUL OR

SPIRIT OVER MIND."
THIS IS DONE BY USING PSI ENERGY, FOR PSI ENERGY IS ALL ENERGY USED AS ONE PSYCHIC ENERGY OR SPIRITUAL ENERGY.
SO A PSI BEING HAS A 'FEELING MIND.' THIS IS CALLED IN THE SCIENCE OF METAPSYCHICS THE PSYCHIC MIND.
THIS IS THE MIND CONNECTED TO THE HEART OF THE UNIVERSE OR THE MASTERMIND.
NO ONE HAS ANY MORE ENERGY THAN ANYONE ELSE SO NO ONE IS GREATER OR BETTER THAN ANYONE ELSE HOWEVER ONE CAN HAVE A BETTER POSITION AND USE OF PSI ENERGY.
YET, THE FACT REMAINS THAT WE ALL HAVE THE SAME AMOUNT OF ENERGY AND THEREFORE THERE IS NO SUCH THING AS 'EXTRASENSORY PERCEPTION'.
WHAT ESP REALLY IS, IS "EXTENDED SENSORY PERCEPTION".

YOUR SENSES ARE SHARPER, YOUR FOOD TASTES BETTER, YOU BREATHE EASIER AND BETTER, AND YOUR LIFE IS MORE ALIVE!

HAVING REAL ENERGY YOU CAN GIVE LIFE TO ANYONE OR TAKE IT AWAY.

WHAT YOU CANNOT DO IS DESTROY ANYONE COMPLETELY.

KRISHNA DOES NOT KILL ANYONE BUT PURIFIES THEM, FOR EVERYONE, EVERYTHING, AND EVERY PLACE IS PURIFIED BY KRISHNA SOONER OR LATER.

I PURIFY MY MATTER AND MASS, MY ENERGY, MY TIME AND SPACE.

I PURIFY KRISHNA AND MY FAMILY AND FRIENDS.

I PURIFY MYSELF.

THIS IS TRANSCENDENTAL CONSCIOUSNESS.

TK!

NOW THE CREATIONS BEGIN AND NOT MUCH MORE I CAN TELL YOU.
FOR YOU ARE ON YOUR OWN NOW.
I HAVE GIVEN YOU ALL I KNOW AND YOU NOW CAN CREATE YOUR OWN KNOWLEDGE AND SHOULD.
NO ONE KNOWS YOU BETTER THAN YOU KNOW YOURSELF BUT KRISHNA.
THIS IS WHY YOU CHANT KRISHNA, FOR KRISHNA IS YOUR ETERNAL SELF AND THE ETERNAL GOD OF ALL.
AND NO ONE IS GREATER OR EQUAL TO KRISHNA THOUGH WE ALL SHARE IN HIS VAST QUANTITIES OF ENERGIES.
YOU CAME FROM BEING THE 'CREATED' TO BEING THE CO-CREATOR.
THE PSI BEING IS COMPLETE AND SATISFIED WITHIN HIMSELF BUT HE DOES STRIVE TO DO BETTER AND THEN HIS BEST AT ALL HE DOES.
SO ONE DOES GOOD AND WHEN ONE IS

GOOD, ONE DOES BETTER, AND WHEN ONE DOES BETTER, ONE DOES HIS BEST ACCORDING TO WHAT HE KNOWS.

THIS CREATES 'NO KARMA' AND 'NO EXTERNAL JUDGMENT' THOUGH 'SUGGESTIONS' SHOULD BE NOTED.

THE PSI BEING IS THE MASTER OF HIS OWN SPACE, TIME, ENERGY, AND MASS/MATTER.

HE DOES NOT STEAL, KILL, OR DEAL IN DECEIT ESPECIALLY IN BUSINESS AND INVESTMENT DEALS.

HE RESPECTS ALL RELATIONSHIPS AND DOES NOT BREAK HIS MARRIAGE VOWS BY 'CHEATING' ON HIS SPOUSE, OR BREAK HIS WORD {as in a vow} FOR ANY REASON.

THIS IS NOT A THING OF 'FORGIVE ME, FOR I HAVE SINNED!' IT IS A THING OF 'I WILL NOT DO THIS, NO MATTER WHAT COMES, PERIOD!'

THE FIFTH LAW CANNOT BE ENFORCED

AND BE THE FIFTH LAW.

WE CALL IT THE LAW OF ETHICS.

THE LAW OF ETHICS STATES THAT 'THESE ARE LAWS I KEEP FOR THE SAKE OF MY FREEDOM, SAFETY, AND SANITY.'

IN OTHER WORDS, THESE ARE THE LAWS YOU USE TO RULE YOURSELF SO THAT OTHERS WILL NOT AND CANNOT RULE YOU OR RULE OVER YOU.

THE LAWS OF ETHICS MAKE YOU ABSOLUTELY SELF-DETERMINED WITH ENTHUSIASM!

IT ALSO COULD BE STATED AS: 'HE WHO RULES HIMSELF, WILL RULE OTHERS.'

SOUL CONTROL IS THE NEW WAY TO CONTROL AND DISCIPLINE YOURSELF SO THAT YOU CAN SET GOALS AND ACHIEVE THEM WITH EASE AND SIMPLICITY.

'SPIRIT {Soul} OVER MIND' IS GREATER THAN

'MIND OVER MATTER' AS

'MIND OVER MATTER {Might}'
IS GREATER THAN
'MIGHT OVER MATTER.'
EVEN AS THE ETERNAL IS GREATER THAN
THE INFINITE AS THE INFINITE IS
GREATER THAN THE FINITE.
"HARE KRISHNA, HARE KRISHNA,
KRISHNA, KRISHNA, HARE, HARE!
HARE RAMA, HARE RAMA, RAMA, RAMA,
HARE, HARE!"
THIS IS THE HIGHEST TRANSCENDENTAL
SACRIFICE AND THEREFORE WAY!
A PSI BEING USES IT REGULARLY.

I woke up!
It's about 12:19 am.
I wake up at different times, letting my body dictate when I eat, rest, go to the bathroom, move, do an activity, meditate, envision, or just sit there and BE.
Yes, a PSI BEING can just sit there and BE.

He does not have to DO anything.
Not think, or act, or create, just BE.
"YOU ARE THE WHOLE EXPRESSION OF THE UNIVERSE IN HUMAN FORM.
YOU ARE AN ETERNAL SOUL AND THEREFORE HAVE INFINITE ABILITIES AND ATTRIBUTES.
YOU ARE NOT THE WEAK BUT THE STRONG.
YOU ARE NOT THE FEARFUL BUT THE FEARED.
PEOPLE CANNOT SEE YOU.
PEOPLE CANNOT HELP YOU.
THIS WORLD IS NOT DESIGNED FOR YOU.
YOU CREATE OR CO-CREATE YOUR OWN WORLD, YOUR OWN UNIVERSE.
YOU NEED NO EXTERNAL LIGHT THERE, YOUR UNIVERSE IS SELF-LUMINOUS!
I SEEK THE POWER AND JOY WITHIN ME NOT WITHOUT.
I HAVE THE STRENGTH OF A THOUSAND

MEN, THE ENERGY OF AN INFINITE SOURCE.

AUM= I AM TRANSCENDENTAL.

SAT=I AM IN MY ETERNAL HEAVEN.

TAT=I AM A CO-CREATOR OF ALL {GOD}.

MY ATMOSPHERE IS PURIFIED.

MY ENVIRONMENT, BODY, AND MIND ARE PURIFIED.

I AM KRISHNA CONSCIOUSNESS OR TRANSCENDENTAL CONSCIOUS.

TIME IS ON MY SIDE.

SPACE IN IN MY HANDS.

ENERGY IS IN ABUNDANT SUPPLY.

MATTER IS ENOUGH FOR TODAY.

SUFFICIENT IS THE 'EVIL' OF TODAY.

THERE IS NO LACK OR SLACK OF LOVE, OR JOY, WISDOM, GOD!

THE 'VOID' IS AN ILLUSION, WE HAVE NO NEED OF ANYTHING.

I LEAVE MY PHYSICAL BODY AND EMBRACE MY 'LIGHT BODY' OR THE FIELD

OF ELECTROMAGNETIC ENERGY AROUND ME.

I CHANT MY SECRET MANTRA AND FEEL THE PSI ENERGY WITHIN ME.

I AM CREATING MYSELF. I AM RE-CREATING MY ATMOSPHERE AND ATTITUDE.

NOTHING IS FIX. NOTHING IS STATIC BUT KRISHNA.

I BREATHE IN INFINITE LOVE, JOY, AND FREEDOM {POWER}.

I BREATHE OUT RELAXATION AND PEACE.

MY SLAVE-mind STIRRING THE IMAGINARY ENERGY.

TWISTING IT SO I THINK SOMEONE OR SOMETHING DONE SOMETHING TO ME I MUST SEEK REVENGE FOR.

THE EVIL DONE TO ME, WHO ADDRESSES THAT?

IT IS ALWAYS MY FAULT! I MAKE MYSELF THE VICTIM REGARDLESS OF WHO

STARTED THE FOUL ACT!

MEMORIES OF PAIN! MEMORIES OF HATE AND CONFUSION.

I TAKE LEAVE AND GO LIGHTYEARS AWAY.

THESE DEAD DREAMS OF YESTERDAY!

WASH THEM AWAY! WASH THEM AWAY!

DWELL ON THIS AND DWELL ON THAT!

NO!

IT IS A NEW DAY AND A NEW DAY REQUIRES NEW THOUGHTS!

AND THE WORLD IS CHANGING.

OH YEAH OF LITTLE FAITH!?

HUH! WHAT FAITH IS NEEDED?

OH YEAH OF GREAT CREATIVE POWER!

THE WAY IN IS THE WAY THROUGH.

THE PSI BEING IN HEAVEN CAN STAY IN HEAVEN

THOUGH IT IS FAVORED THAT ONE ASCENDS TO EARTH TO HELP OTHER DEVOTEES.

BUT THESE MAY BE FEW AND FAR

BETWEEN.

YOU THEREFORE MUST CO-CREATE DEVOTEES.

YOU MUST REMIND THEM WHAT THEY TRULY ARE AND LET THEM DISCOVER WHAT THEY CAN TRULY DO.

AS A PSI BEING YOU ARE A SPIRITUAL MASTER SO DO NOT EXPECT MUCH HELP FROM OTHERS, THEY SIMPLY HAVE NO IDEA OF THE KNOWLEDGE YOU POSSESS AND MUST RELY ON THE KNOWLEDGE THEY ALREADY HAVE.

LEARN TO GIVE YOURSELF WHAT YOU NEED AND WANT AND DESIRE.

A RICH MAN CAN BE SPIRITUAL AS EASILY AS A POOR MAN CANNOT.

IN MOST CASES, IT IS THE POOR MAN WHO IS NOT SPIRITUAL, FOR SPIRITUALITY REQUIRES THAT ONE TAKES CARE OF ONE's ENVIRONMENT, BODY, AND MIND HERE AND NOW.

THE BEGINNING AND THE ENDING IS EASY IT IS THE MIDDLE PART THAT YOU MUST WORK ON AND THAT IS GREAT FOR THE MORE YOU WORK ON IT THE MORE IT TURNS INTO 'PLAY', "PLAY-WORK'.

PLAY-WORK IS A TERM FOR A PSI BEING FOR HE DOES NOT DIRECTLY 'WORK' ON ANYTHING. IT IS MORE A 'PLAY-WORK'. HE PLAYS AT WORKING MEANING THAT HE FINDS THE DEEPER MEANING AND PURPOSE IN HIS ACTIVITIES ARE 'PUTS' IT THERE THROUGH KRISHNA CONSCIOUSNESS.

THEREFORE YOU HAVE PERSONAL AND IMPERSONAL PSI BEINGS.

ONE IS NO DIFFERENT FROM THE OTHER EXCEPT THE PSI BEING KNOWS AND UNDERSTANDS HE OR SHE CAN TAKE A TRANSCENDENTAL POSITION TO EVERYTHING, EVERYONE, AND EVERY PLACE.

THIS WOULD BE TRANSCENDENTAL
CONSCIOUSNESS.
THE UNIVERSAL CONSCIOUSNESS IS
KNOWN TO MANKIND AS:
AUM-TAT-SAT.
NARAYANA
IS FOR SPIRITUAL AND PSI BEINGS.
AND
HARE KRISHNA HARE KRISHNA, KRISHNA,
KRISHNA, HARE, HARE,
HARE RAMA. HARE RAMA, RAMA RAMA,
HARE HARE!
THAT IS IT.
THE REST IS PSI PRECEPTS.
YOU CAN TAKE A MAN OR WOMAN TO
THE PATH TO HEAVEN BUT YOU CANNOT
MAKE THEM WALK IT AND YOU CANNOT
WALK IT FOR THEM.
THIS IS YOUR SAVING GRACE!
YOUR ONLY TRUE AND REAL JOB IS
YOURSELF:

KRISHNA.

Is it me? Or is it that when you keep a daily Journal you see how fast time flies? It seems like I just wrote in my Journal a few hours ago and now it is a new day!

I sit here and try not to judge myself. What is right with me? What makes me happy? How can I use this time to my advantage and forget about those who cannot face the TRUTH?

I go into my TRANSCENDENTAL MEDITATION.

I AM TRANSCENDENTAL CONSCIOUS.

I AM THREE INCHES ABOVE MY HEAD AND MY BODY IS FEELING WAY BETTER THAN IT DID A WEEK AGO.

I AM LISTENING TO MY FIRST ALBUM AND WATCHING THIS HORROR PICTURE ON TV.

HOW USELESS I HAVE GOTTEN. WANT TO

LISTEN TO HEALING MUSIC, THE PARTA IS TARDY AND I'M NOT HEARTY.

I HAVE FREE WILL TODAY? NO. I HAVE A FREE CHOICE!

LET COME WHAT WILL! I CANNOT FOLLOW OR BE PART OF THE HERD MIND.

I CONNECT TO THE PSI ENERGY WITHIN ME.

I BREATHE.

I EXERCISE.

I AM ALIVE!

I DO A PSI PROCESS.

I THINK HOW CAN I REALLY ENJOY MYSELF?

HOW CAN I USE THIS TIME SO THAT I AM AHEAD OF THE GAME?

I FIND I AM IN A CLASS ALL MY OWN. I GET MY WORK DONE THOUGH IT IS SLOW IN COMING.

HOW MANY OTHER AUTHORS STRUGGLE WITH THE MIDDLE?

THE MIDDLE IS THE HARDEST PART!

FOR YOU MUST CARRY THEM OVER THE RIVER,

FORGET TIME, SPACE, AND ENERGY.

IT IS NOT A MATTER OF HOW YOU FEEL!

YOU WILL GET NOTHING DONE AT ALL!

YOU GOT THE HELP YOU NEED WHAT IS YOUR PROBLEM?

YOURSELF? JESUS? YOUR ETERNAL MIND?

YES, YOU ARE THE WITNESS OF ALL YOU DO AND DON'T DO.

SO ARE YOU TRYING TO BE IMPERFECT OR PERFECT?

YOUR SOUL CANNOT GET ANY BETTER, ONLY YOUR MIND, BODY, AND ENVIRONMENT!

YOU PUT ORDER TO CHAOS.

THE MIND IS A BALL OF CONFUSION.

WE USE A STABILIZING AXIOM.

KRISHNA.

A PSI BEING IS A WILLING CAUSE OVER

HIS OWN SPACE, TIME, ENERGY, MASS, AND MATTER.
MASS IS THE 'SOLID' ENERGY OF THE MIND.
IT IS WHAT MENTAL PICTURES ARE MADE OF.
THE DAY IS SNEAKING ON BY.
I AM AT PEACE AND WATCHING IMPACT THEORY ON YOUTUBE.
MY BODY IS MORE BALANCED AND I HAVE HARDLY ANY LOWER BACK PAIN!
IT IS STILL UNCOMFORTABLE BUT A LOT BETTER THAN IT EVER HAS BEEN.
I GO INTO SAMADHI.
THE DAY IS AS QUIET AS I AM.
PUT ON SOME MUSIC TO MAKE IT BETTER AS I WRITE.
MAKING THIS A FUN AND HAPPY EXPERIENCE!
WELL, AN EXPERIENCE!
I THINK, WHAT ARE MY CHOICES TODAY?

WHAT ARE MY LIMITS?
SO I CAN EXCEED THEM!
YOU DO NOT HAVE TO TEACH AS I DO.
YOU ARE FREE TO BE AS YOU PLEASE.
THE GOAL IS A PLEASANT LIFE.
A HAPPY LIFE AND JOYFUL LIFE!
I DO NOT FEEL GOD ALL THE TIME
THE WAY I WANT TO!
GOD CHANGES THE WAY I FEEL AND I
FEEL KRISHNA.
I AM READY TO DO MUSIC.
ALWAYS DREAMING OF MY AUDIENCE.
THE PEOPLE WHO LOVE AND APPRECIATE
ME,
THE PEOPLE I LOVE AND APPRECIATE!
ALL THESE HIDDEN FEARS COME TO
LIGHT!
WHAT AM I AFRAID TO WRITE?
I WAS THIS AND I WAS THAT
CAN'T FORGIVE THIS AND WON'T
FORGIVE THAT!

GOTTA LIVE WITH THIS THE REST OF MY LIFE!
I HAVE ESP AND KNOW AND UNDERSTAND.
YOU'LL FEEL GOD'S POWER WHEN YOU TAKE MY HAND,
I AM A PSI BEING, MORE THAN JUST A MAN.
I AM A PSI BEING AND KNOW HOW TO TRANSCEND!
YOU CAN JUST SIT AND YOU WILL STAY BALANCED AND CAN FOCUS EVER EASY.
I COULD SIT ON THE MOON ALL DAY, DRINKING MY FAVORITE WINE, OR SMOKING MY FAVORITE BLUNT.
CELEBRATE YOUR LIFE AND YOU DO GOD'S WILL.
I PURIFY MY LIFE, MY SPACE, TIME, ENERGY, BODY, AND ENVIRONMENT.
I CONCENTRATE ON MY LIGHT BODY, MY ELECTROMAGNETIC BODY.

I DO TRANSCENDENTAL MEDITATION
WATCHING SISTER ROSETTA THARPE,
GODMOTHER OF SOUL.
AUM-TAT-SAT.
I AM IN MY OWN ENERGY.
I DIE TO MYSELF.
PEACE WITHIN AND WITHOUT.
LOVE IS ALL ABOUT!
I LET GO OF ALL PAST AND FUTURE FEAR!
I BECOME CREATIVE AND THINK OF
WHAT I WILL PREPARE FOR DINNER!
WHAT CAN I DO FOR FUN TONIGHT?
GOT TO GET OUT OF MY OLD COMFORT
ZONE.
I WALK ON THE MOON!
WHERE DID ALL THE WORDS GO?
YOU HAVE THE LAW OF ETHICS NO ONE
CAN ENFORCE ANY LAW ON YOU AND YOU
ARE NOT REQUIRED TO FOLLOW ANY
LAWS BUT YOUR OWN, BUT ALL ACTIONS
HAVE THEIR PROPER CONSEQUENCES.

It is like this day came and went already and I have just barely begun to write! I slowed way down because I want a perfect print so it needs little editing if any at all!

The only thing that you can really change in this world is YOURSELF and that is enough. But remember, the self is also part and parcel of the UNIVERSE, Solar System, Earth, World, Environment, and your immediate Environment.

YOU CANNOT DIVORCE THE PHYSICAL BODY FROM THE PHYSICAL UNIVERSE NO MORE THAN YOU CAN SEPARATE THE MIND FROM THE BODY.

AND THE MIND CANNOT BE SEPARATE FROM THE SOUL AS THE SOUL CANNOT BE SEPARATE FROM THE SUPREME SOUL.

THERE IS NO SUCH THING AS SOMETHING THAT IS NOT PHYSICAL.

ENERGY IS PHYSICAL AND IF SOMETHING

IS NOT PHYSICAL, THEN IT IS THE PROPERTIES OF THE PHYSICAL ARE ENERGY.

MOST OF LIFE LIVES IN PRAKRTI, THE PSI BEING IS THE CONSCIOUSNESS OF DAIVI PRAKRITI WHICH IS A PARADISE ON EARTH.

FROM DAIVI PRAKTRI IS BRAHMA AND ON UP TILL YOU GET TO THE ETERNALLY HEAVENLY BODY AND ABODE.

THIS IS HOME, YOUR HOME, EVERYBODY IS GETTING TO OR WANTS TO GET TO.

A PERSON WANTS TO GO TO A PLACE WHERE HE IS LOVED THE MOST AND THAT WOULD BE WITHIN YOURSELF!

I BREATHE IN DEEPLY AND EXHALE COMPLETELY.

MY BODY IS VERY TOLERABLE TODAY I AM GRATEFUL FOR THAT!

NEVER WILL HAVE ALL THE ANSWERS AND NEVER WILL KNOW ALL THE KINDS

OF DISEASES THAT CAN INFILTRATE THE BODY.

I AM AT SUPER PEACE WITH MYSELF.

I CAN SEE AN ACTION FOR ACTION AND LEAVE MY JUDGMENT OUT OF IT.

HOW WE WERE PROGRAMMED TO MAKE MISTAKES! SAY NOTHING OF SIN.

SO NOW SIN AND KNOW IT.

DO WHAT YOU ARE AFRAID TO DO AND DO IT GLADLY WITH COURAGE.

ALL ACTIONS HAVE CONSEQUENCES, DO YOUR DUTY AND LET KARMA TAKE CARE OF ITS OWN.

ALL IS QUIET AND THAT IS KRISHNA!

I AM IN KRISHNA CONSCIOUSNESS OR TRANSCENDENTAL CONSCIOUSNESS.

I FINISHED ONE SONG, 'SIC AND TIRED' AND GIVE IT A THUMBS UP BUT NO CIGAR.

I AM NOT SICK AND TIRED AS I USED TO BE.

MY BACK IS TWISTING TO THE LEFT, MY LEFT LEG IS ALL TIGHT AND MY DICK SEEMS STUCK ON THE LEFT SIDE! WTF? NOT SURE WHAT THE HELL IS GOING ON BUT MY SAVING GRACE IS THAT I FEEL MORE HEALTHY THAN I DID THREE MONTHS AGO.

AND I HAD BACK PAIN YESTERDAY THAT IS NOT HERE TODAY!

MY BACK IS STILL WEAK AND DOING STRANGE THINGS, I AM MORE STABLE AND CAN GET THINGS DONE IN STRIDE. GOT SOME ANTS IN MY APARTMENT, AT LEAST THEY ARE NOT ROACHES, AND HAVE NOT SEEN ONE IN YEARS AND I MEAN YEARS! KNOCK ON WOOD!

I WAS KICKED OUT OF EVERY MENTAL HEALTH PROGRAM. I COULD NOT ADJUST TO BULLSHIT. I WAS BORED AND THEY ACTUALLY WENT IN REVERSE TO ANY NATURAL TALENT I HAD!

AND THIS IS THE PROBLEM FACING MYSELF AND BLACK CREATORS TOMORROW.

THERE IS NO SUITABLE OUTLET FOR A MAN OF ART. A MAN OF SELF-CREATING AND RE-CREATING.

A PSI BEING IS A CO-CREATOR AND CAN 'ACT' AS THE 'CREATED'.

HE IS A CAUSE AND CAN 'ACT' AS THE FIRST CAUSE AND CAUSE UNDER THE CAUSE OF CAUSES THAT NEEDS NO CAUSE.

THE PSI BEING REALIZES THAT THERE ARE INVISIBLE VIBRATIONS OR FREQUENCIES THAT GOVERN THE UNIVERSE, EARTH, WORLD, MY IMMEDIATE ENVIRONMENT, BODY, AND MIND.

YOU ARE ALREADY WHAT YOU WANT TO BE!

YOU ARE DOING WHAT YOU NEED TO DO:

BREATHE-STRETCH{FLEX}, BUILD YOUR INNER STRENGTH {Light Body}.

YOU ARE NOT AFRAID TO CARE.

YOU ARE NOT AFRAID TO BE HERE.

YOU ARE NOT AFRAID OF YOURSELF OR YOUR SPIRITUAL LIFE!

IT IS THE FACT YOU CAN CHANGE YOUR CONSIDERATIONS AND THE CONSIDERATIONS OF OTHERS THAT THERE CAN BE SANITY.

REMEMBER, THERE ARE THREE UNIVERSES,

YOUR UNIVERSE,

THE UNIVERSE WE ALL AGREE OR DISAGREE ON,

AND THE OTHER FELLOW's UNIVERSE.

GRANT SPACE AND TIME TO YOURSELF AND OTHERS.

YOU CAN APPROACH LIFE IN AN IMPERSONAL, PERSONAL, OR PRIVATE WAY, THE CHOICE IS YOURS.

YOU CAN OPERATE IN ANOTHER UNIVERSE AND HAVE COMPLETE CONTROL OF YOUR ENVIRONMENT, BODY, AND MIND

THIS IS CALLED "SOUL CONTROL."

THE HIGHEST REASONING AND LOGIC SYSTEM:

"SPIRIT OVER MIND."

GOD MAKES IT EASY FOR ANYONE TO KNOW, UNDERSTAND AND USE HIS INFERIOR ENERGIES TO FIND HIM AND LIVE YOUR TRANSCENDENTAL LIFE!

GOD IS REAL WHEN YOU SEEK THE LIVING GOD AND NOT THE DEAD ONE.

I woke up shitty.

That is all that is too it.

I have a back problem or skull problem or groin problem.

I have one real friend. And the rest of the day I am alone.

Used to go see my Stepmother and sister but they did not welcome me and looked as if they didn't even want me over there.
I don't like Christianity and I don't like Islam either.
Black people do not even TALK to each other less known come together to do something about the black problem in America.
Black Chrisitans are like in a Hamster Cage. They got their little house and car and food on the table for the next generation but care less about what is going on TODAy, right now! You cannot even get these stupid black women to listen. They are so programmed by the white man.
I wake up and I want to cry, yes! I worked so hard to be a decent person, a happy person and still no one sees me, talks to me, asks me what is new?
And they say I am to blame.

Fuck it!
I go to the moon.
I find love fro, my brother, my close friend-brother.
I put this happy scene before me!
I feel pretty good now.
The only thing I can change is myself.
Let me do that.
Help me to forget what has happened and how I am not responsible for it.
I am neutral.
We all die alone.
I seek nothing today but to finish this page.
It is hard because I really don't know what to write.
I listen to music and know I will see people but people do not see me.
I died and no one knew it, didn't anybody really care?
I listen to music.
I am not hungry but at least I can sit up and

do stuff on the computer.
Have no wish to go outside.
AUM TAT SAT
I am getting my POWER from within!
I am thinking and feeling what is GOOD for myself.
I have myself! I have my own ENERGY and LIFE!
Enter into God's abode.
I am here with the Witness, with God the Eternal.
In this ENERGY, this PSI ENERGY I am.
I feel I can make it but it makes me so mad the bitch dont understand.
Trying too much to be like white people, that is what Christianity is all about.
My neck is constantly hurting me. I have to lay down every two or three hours.
My Energy is weak, I can only do so much, move too long and I am tired as a dog.
Feel like my life is coming to an end and I

just try to forget the black women in my life, doing nothing for me but cause me pain and misery.

I really wish they would all die.

I have to do what I have to do and it makes me so mad that even my family does not know what is going on.

I got to do something.

But what can I do?

I made friends.

Huh, why can' anyone talk to me?

I am trying to finish my page out.

I really don't care. I've done so much and never got credit or any thank you for it a 'hey, Bruce, this is very nice.

No, I never had any encouragement and when I wrote my first book, it changed nothing. No one in my family was even interested.

What should I do?

Your own family does not really care for

you, how can I stop caring for them?
It does nothing but bring me down, how can
I get rid of them? How can I make them
sorry for what they did?
Show them Satan and bless them they
needed.
I sit here and think I have the day before me
and I feel OK.
But there is no life, no soul, no thought.
I am dry as ice.
I will not stay this way.
I did nothing wrong but speak the truth like
I always did.
People will reject me for the rest of my life.
I don't care.
I give in.
Except for my true friends they all are
blessed.
Blessed of Satan.
That is the best I can do.
I walk alone and will do as I please.

These dead Christian Zombies have nothing on me!
Let them be a stupid sinner and live a life of illusion.
I don't care anymore.
If I ever did in the first place!

OK!
Another day. 24 hours. How can I waste it?
I try to do some projects. My legacy you might call it, but who will see it or even read it?
Who am I that anyone should take interest in?
And how is it that I take interest and get nothing in return?
Work all my life to be here and now that I am here I am going to
ENJOY my life or your LIFE, it is all the same to me!
Yesterday was a day of pure hell!

My back was cramping again so hard to move and moving in great pain! How can I explain?

But I am AUM TAT SAT!

What a beautiful FEELING despite my body trying to feel pain and discomfort.

HOW IS IT I CONDEMN MYSELF FOR THE LEAST THING?

SITTING HERE I SEE WHERE THE FEAR IS AND HOW I THINK IT IS OVER WHEN IT IS ONLY BEGINNING.

I AM APATHETIC, UNCARING AND WONDERING WHAT DO I REALLY LOVE?

OR DOES IT MATTER WITH KRISHNA?

EVERYTHING IS HIS WILL!

I DISAGREE WITH MY ENVIRONMENT, MY BODY, MY MIND, AND MY SENSES.

I FLOW ENERGY. I MOVE OUTSIDE MY BODY INTO MY IMMEDIATE ENVIRONMENT.

MY MIND IS ADJUSTING ITSELF FOR FUN

AND PROFIT.

WORK-PLAY IS WHAT THE PSI BEING DOES. EVERYTHING IS SPIRIT IS A 'PLAY ON WORDS, THOUGHTS, AND FEELINGS.

NOW YOU HAVE CONTROL OVER YOUR FALSE E-GO AND E-MOTIONS.

I AM YOUR FATHER AND MOTHER LET US GO TO OUR ETERNAL PEACE!

I AM YOUR GRANDFATHER AND GRANDMOTHER LET US GO TO OUR ETERNAL PEACE!

I AM YOUR GREAT-GREAT GRANDFATHER AND MOTHER LET US GO TO OUR ETERNAL PEACE!

I AM YOUR BEST OF FRIENDS, EVEN YOUR SPIRITUAL LOVER!

MAKE OF ME WHAT YOU WILL AND I WILL BE STILL!

YOU CAN'T CHANGE ME, BUT I CAN CHANGE YOU.

I CAN CHANGE ANYTHING.

YOU ARE LIKE ME AND I AM LIKE YOU!
YOU CAN'T OUTGROW ME, BUT I CAN OUTGROW YOU!
WE TRAVEL THIS ROAD TOGETHER!
GOOD OR BAD WEATHER!
I SEEK YOU OUT DAILY!
YOUR COMPANY I ENJOY ALWAYS!
THE PEACE, THE COMFORT, THE LOVE AND JOY!
THE POWER AND WISDOM, PROSPERITY AND JOVE!
YOU ARE UNDER MY PROTECTION I AM WITH YOU ALWAYS.
THOUGH I APPEAR AND DISAPPEAR, I DO SO WITH YOU.
WE ARE ONE BUT NOT EQUAL. WE ARE ONE AND COMPLETELY THE SAME BUT YOU ARE THE INFINITE AND I AM THE ETERNAL.
YOU CAN BE ANYTHING.
YOU CAN DO ANYTHING.

YOU CAN HAVE ANYTHING.
YOU CAN CREATE AND DUPLICATE ANYTHING.
ANY SPACE.
ANY TIME.
ANY ENERGY FORCE OR FORM.
AND MATTER, MASS, BLACK MATTER OR ANTI-MATER.
YOU CAN LIVE IN DEGREES OR LIVE TO THE ULTIMATE.
YOU CAN HAVE AN EFFECT AND BE THE EFFECT OF IT!
YOU CAN DIE WHEN YOU LIKE AND REVIVE ANY MOMENT.
YOU ARE THE MASTER OF YOUR E-GO AND E-MOTIONS.
CAST THIS FEELING HERE OR THERE OR EVERYWHERE.
BRING THIS FEELING IN, DEEP, AND STIRRING.
THE ONLY THING YOU CRIED ABOUT WAS

FEAR AND DEATH.

FEAR OF DEATH.

DEATH OF FEAR!

TAKE YOUR FEAR WITH YOU!

FEAR, DEATH, LOSS, AND SORROW!

THE ILLUSION THAT SPIRIT DOES NOT POSSESS EVERYTHING, EVEN THAT WHICH IS NOT CREATED YET.

THREE FOURTH OF THE WORLD IS NOT SEEN.

OPEN YOUR SPIRITUAL EYES!

YOU'RE IN FOR A BIG SURPRIZE!

DO WHAT YOU NEED TO DO

TO DO WHAT YOU HAVE TO DO

TO DO WHAT YOU WANT TO DO.

THIS CONFIDENTIAL KNOWLEDGE IS TRANSCENDENTAL KNOWLEDGE, THE WORLD KNOWS NOTHING OF.

BUT THERE ARE 'LIGHTS' IN THE WORLD AS WELL AS 'THE DARKNESS'.

FRET NOT. WORRY NOT. HESITATE NOT.

YOU ARE AN ORIGINAL SON OF GOD.
BEWARE THE HOLY BOOKS THEY HAVE BEEN ALTERED TO SERVE THE POWERS THAT BE INSTEAD OF YOU AS IN INDIVIDUAL.
AND AS AN INDIVIDUAL YOU MUST TAKE CARE OF YOUR OWN SENSES, MIND, BODY, AND IMMEDIATE ENVIRONMENT. THIS IS YOUR FIELD OF POWER. YOUR SPHERE OF CONTROL.
SOUL CONTROL!
IT IS A GIFT OF GOD TO BE ALONE, IN SOLITUDE FOR ONLY THEN CAN YOU FIND THE REAL AND TRUE HEAVEN WITHIN!
I WOULD NOT HAVE WRITTEN SO MANY WORDS TO YOU IF THIS WERE NOT SO.
AND I AM LONG GONE.
THAT IS AS BRUCE LEMAR PETERSON.
I AM THE PRIME PSI BEING.
THIS IS MY SACRIFICE. MY LIFE'S WORK

TO YOU.
YOU HAVE REACHED THE ULTIMATE OF YOUR LIFE.
IT CANNOT GET ANY BETTER THAN THIS EXCEPT AS YOU DO IT IN KRISHNA CONSCIOUSNESS OR
TK!
THE WORLD NEEDS PEOPLE LIKE YOU!
DON'T LET THEM DOWN LIKE THEY DID YOU.

Good Morning!
What a day this is!}
The day itself is so kool!
I only think about my body, the pain everyday I respond to!
The isolation from family and not ever given a chance to TALK TOO!
I find it hard to forgive ignorant fools!
Looking at me like I am the anti-Christ, not knowing how BOX IN we are in!

THE LOVE OF GOD IS IN YOU. ALSO HIS HEALTH, WEALTH, LOVE, JOY, POWER, WISDOM, ABUNDANCE, PROSPERITY, AND PEACE TO NAME A FEW OF HIS ATTRIBUTES.

GOD IS ETERNAL SO NO ONE, {NOT EVEN JESUS OR MOSES OR HORUS OR MOHAMMAD} CAN BE ONE WITH GOD. EVERYONE IS PART AND PARCEL OF GOD AND HAVE THEIR JOB OR WORK TO DO. WE ALL SERVE A PART IN THE WHOLE SCHEME OF THINGS AS INDIVIDUALS. THERE IS ONLY THE GOD OF SELF! ALL OTHER GODS ARE FALSE OR OF THE FALSE EGO.

YOU HAVE GONE FROM THE CREATED TO THE CO-CREATOR.

MAN THINKS HE CREATES AND GOD CREATED OUT OF NOTHING , BUT MAN AND WOMAN CAN ONLY REALLY RE-CREATE WHAT IS ALREADY HERE, AND

WHAT IS ALREADY HERE IS THE FORM OF GOD AS THE LIVING UNIVERSE.

SO, A RELIGIOUS PERSON LIVES IN A 'DEAD WORLD OF DEAD IMAGES AND WORDS', WHILE THE SPIRITUAL PERSON OR PSI BEING LIVES IN A WORLD OF ETERNAL BLISS {Joy of God}, KNOWLEDGE {Spiritual and Transcendental Knowledge from God or Co-creator of God}, AND LIFE {Love of God}.

A PERSON ON THE FIRST LEVEL OF METAPSYCHICS IS CALLED A SON OF GOD.

THIS IS A READER. A PERSON DOES NOT HAVE TO PRACTICE METAPSYCHICS AT THIS TIME AND JUST LISTENING TO METAPSYCHICS WILL MAKE ONE BETTER, HOWEVER, WE CANNOT SAY HE OR SHE WILL REMAIN THAT WAY WITHOUT FURTHER STUDY.

EVERYONE IS POTENTIALLY A SON OF

GOD AND THIS IS A TITLE AND POSITION SO IT REFERS TO WOMEN AND CHILDREN ALSO.

A SON OF GOD IS A PERSON WHO KNOWS HE IS NOT HIS OR HERS ENVIRONMENT/PHYSICAL BODY.

THIS IS THE TRUE MEANING OF THE LOGIC SYSTEM:

"MIND OVER MATTER."

WHEN ONE COMPLETES METAPSYCHICS HE IS IN POSSESSION OF SOUL CONTROL, or "SPIRIT OVER MIND", a new logic/fantasy reason system of thinking.

TODAY IS GETTING BETTER AND THEN I THINK ABOUT MY PAINFUL BODY.

MOVING THIS WAY AND THEN THAT WAY, THE PAIN IS GONE, NO MY MIND IS OFF OF IT!

EVERYDAY THIS PAIN, I TRANSCEND.

I MUST DO SOMETHING ABOUT IT!
HAVE NO IDEA WHERE IT IS COMING FROM!
CAN'T LET IT KEEP THE FUN OUT!
HOW DETERMINED I AM TO HAVE FUN TO LIVE A LIFE COMPLETE TO THE BEST OF MY ABILITIES.
I LOOK AROUND AND I AM HAPPIER!
I AM PART OF THIS EARTH, PART OF THIS WORLD!
IT WAS CREATED FOR MY PLEASURE, MY END!
WHAT IS THIS HATE OF THE WORLD!
HATE OF MY MIND!
HATE OF MY PHYSICAL BODY!
HOW DARE YOU MAKE IT FEEL GOOD AND NOT LET IT SUFFER!
WHAT IS THIS HATE OF LIFE AND PEOPLE AND SEX!
AS IF I HAVE TO FURTHER THE HUMAN RACE THAT IS OVERPOPULATED

ALREADY.

YOUR PLAN IS COMING AN END!

YOUR THOUSAND YEAR REIGN IS OVER!

WE SEE THE DEVIL FOR WHAT HE IS AND HOW HE WOULD LOVE TO TAKE OVER!

JESUS LOVES ME, I AM HIS BROTHER!

BUT THIS FALSE JESUS IN THE WORLD IS THE DEVIL!

HE WANTS TO TAKE OVER!

I KNOW THE POWER OF GOD IS WITHIN ME.

I AM AN INDIVIDUALIZATION OF THE ONE SUPREME BEING WHO IS ETERNAL BLISS, KNOWLEDGE, AND LIFE.

A FINITE VERSION IS THE CREATED.

THE INFINITE VERSION IS THE CO-CREATOR.

I CO-CREATE WITH GOD.

I CREATE OUT OF GOD EVEN AS GOD CREATED EVERYTHING FOR ME.

I AM A SON OF GOD, AN EXPANSION OF THE LIVING GOD IN THE EARTH AND WORLD.

MEDITATE ON ME!

I AM INDESTRUCTIBLE! I CANNOT DIE FOR I WAS NEVER BORN-THAT IS WHAT IS CALLED ETERNITY AND THERE CAN ONLY BE ONE ETERNITY.

"I Am the ONE that became the MANY!"

Don't be afraid to LOOK! LOOKING is taboo! LOOKING is SIN! LOOKING is KNOWING TO MUCH or not KNOWING AT ALL!

LOOKING is a crime!

The emperor wears no new clothes!

No one KNOWS how near or how far!

YOU ARE A SON OF GOD, MORE THAN HUMAN BUT LESS THAN GOD {The Eternal}.

YOU ARE AN INFINITE CO-CREATOR.

THERE NEED BE NO END TO YOUR CREATIONS.

YOU ARE AS IT IS!

I woke up and I can tell I Have low energy but I feel alright except for my left side and head.

I am not sure what is happening with me.

They say it is arthritis but it could be kidney failure.

But I will be 70 this year so I have to expect some aches and pains.

I feel the Qi and I go into SAMADHI.

AUM TAT SAT!

I FEEL THE GOOD ENERGY AROUND ME.

THE AIR IS PURE, THE WATER IS PURE, THE ENERGY IS PURE AND THE FIRE IS PURE AND THE SPACE IS PURE AND THE MASS AND MATTER IS PURE!

MY TIME IS PURE AND MY BODY AND MIND IS PURE.

I AM SOUL CONTROL.

YOU ARE HERE AND NOW, FREE FROM

ILLUSION AND DELUSIONS, DOUBTS AND FEARS!

KRISHNA IS BEYOND THE DUALITY OF DELUSIONS.

I POSSESS ALL THE GOOD QUALITIES OF ALL THAT IS GOOD WITHIN ME.

PERFECTION IS A ILLUSION, YOU DO THE BEST YOU CAN AND THAT IS ALL THAT CAN BE EXPECTED OF YOU.

YOU HAVE THE DIVINE RIGHT TO LIFE, TO HEALTH, WEALTH, LOVE, JOY, POWER, WISDOM, ABUNDANCE, PROSPERITY, AND PEACE.

YOU HAVE A RIGHT TO FREEDOM.

YOU ARE ETERNAL LIFE IN A FINITE FORM.

YOU ARE THE SOURCE OF EVERYTHING AND SUPPLY ME WITH ALL I NEED INTERNALLY.

YOU ARE LIFE AND NOT DEATH. JOY AND NOT PAIN.

CHAOS AND ORDER OR OF THE SAME THING. THE SAME SOURCE.

CHOOSE YE THIS DAY!

THE FREEDOM I GIVE YOU, USE IT NOT FOR SELF-DESTRUCTION.

CREATE YOUR LIFE, YOUR GOALS, YOUR OWN HEAVEN.

HELL AWAITS EVERYONE.

THE PAIN, THE CONFUSION, THE MISERY AND LOSS.

LET NO LOSS BOTHER YOU, IT WILL BE FILLED AGAIN.

REMEMBER, YOU CAN CREATE AND DUPLICATE.

YOU ARE IN YOUR OWN ENERGY, YOUR OWN SELF!

YOU HAVE YOUR INSTINCTS, REASON, INTUITION, FEELINGS, THOUGHTS, AND SENSES TO PLAY WITH.

YOU CAN CREATE OR DUPLICATE.

A PSI BEING CAN CREATE ANYTHING OUT

OF KRISHNA AND CAN DUPLICATE
ANYTHING OUT OF KRISHNA, INCLUDING
A PORTION OF KRISHNA, HIMSELF.
KRISHNA IS MAN BECAUSE I AM MAN.
KRISHNA CAN BE WOMAN IF YOU ARE
WOMAN.
KRISHNA IS THE GODDESS OF LOVE AND
FORTUNE.
KRISHNA IS YOUR VERY BEST FRIEND.
HE SEES ALL YOU DO AND KNOWS ALL
YOU KNOW AND LOVES AND LIKES YOU AS
YOU ARE, FOR YOU ARE AS HE IS!
YOU ARE A CAUSE AND YOUR SPACE IS
DISTANCE AND THE PERSON YOU ARE
CAUSE TO IS NOW THE EFFECT oF YOUR
CAUSE.
IF THE PERSON DUPLICATE YOUR CAUSE
AND SEND IT BACK TO YOU AS A CAUSE
FROM HIM YOU GOT COMMUNICATIONS.
WHERE YOU NOT TRAINED TO LOVE
THOSE WHO HATE YOU?

YOU ARE NEUTRAL HERE AND NOW, ALL OLD MEMORIES FADE INTO NOTHING.
BE NOT AFRAID TO FIGHT FOR WHAT IS RIGHT.
WE DO NOT ADVOCATE VIOLENCE BUT FOR A GOOD CAUSE VIOLENT MAY BE NEEDED BUT ONLY IN SELF-PROTECTION.
YOU ARE ONE WITH THE WITNESS OF THE EVER CHANGING UNIVERSE.
THOUGH WE AVOID ALL FIGHTING AS MUCH AS POSSIBLE, YET, IT ONE PERSIST IN FIGHTING WE WILL FIGHT AND WIN.
HOW FAST LIFE GOES, AND TIMES FLIES!
TODAY IS A NEW DAY AND I NEVER FINISHED LIVING YESTERDAY!
IS MY WORLD A WORLD OF WORDS IN PLAY?
FOR THE DOUBT AND THE TRUTH OR OF THE SAME SOURCE, TWO SIDES OF THE SAME COIN, EVEN AS CHAOS AND ORDER IS.

A POSITIVE AND A NEGATIVE WILL
NEUTRAL EACH OTHER.
A NEGATIVE AND A NEGATIVE WILL GIVE
A 'POSITIVE NEGATIVE',
AND A POSITIVE AND A POSITIVE WILL
GIVE A 'POSITIVE.'
THE PSI BEING ACTUALLY HAS NO GOOD
OR BAD QUALITIES UNLESS HE
CONSIDERS THAT HE HAS THEM.
IN HIS ORIGINAL FORM AND FORCE HE
DOES NOT FEEL OR THINK, SENSE OR
CREATE.
WHEN ONE IS FREE HE CAN BE ANYBODY,
ANYTHING, OR ANY PLACE.
HE CAN BE AN ACTION OR A NO-ACTION
BEING.
THAT IS HE CAN ACT AS A STATIC OR A
KINETIC AND HE ALSO CAN TRANSCEND.
DON'T FIGHT YOURSELF IN LACK OF
GAME.
LEARN TO TOLERATE OTHER BODIES,

DEAD AND ALIVE.
EVERYBODY HAS A RIGHT TO LIVE
EXCEPT THOSE HOW ILLEGALLY TAKE
THAT RIGHT FROM OTHERS.
REMEMBER, "I AM" IS ALSO "I WILL", AND "I
WILL NOT."
THIS IS THE PAST, PRESENT, AND FUTURE
OF A PSI BEING,
FOR KRISHNA HAS NO PAST OR FUTURE,
HE IS ETERNALLY PRESENT!
BILLIONS OF LIGHTYEARS FROM HERE, I
COME NEAR, I COME HOME.
THEY TELL ME I AM NOT SUPPOSE TO BE
HERE, BUT A PSI BEING IS NEVER ALONE.

It's a new day and I have been chanting and
I have less pain each day. Nothing is perfect!
I GO INTO SAMADHI.
I DO A FEW EXERCISES FOR MY LEGS.
I GO INTO TRANSCENDENTAL
MEDITATION.

SO GLAD THAT GOD GAVE ME THIS
OPPORTUNITY TO CONVERSE WITH YOU.
YOU ARE A PSI BEING, THE MOST
POWERFUL BEING IN ANY UNIVERSE.
BECAUSE YOU DO SOMETHING THAT
PLEASES YOU DOES NOT MEAN THAT THE
WORLD HAS TO FOLLOW SUIT.
KRISHNA FORCES HIMSELF ON NOBODY.
ATTEND TO YOUR OWN DUTY.
HAVE JOY IN WHAT YOU DO EVEN IF YOU
DON'T FEEL IT ALL THE TIME
THE JOY WILL COME IN TIME.
DON'T FEAR! DON'T WORRY! DON'T
HESITATE!
NO ONE CAN HELP YOU; YOU CAN ONLY
HELP OTHERS.
YOU CAN GIVE LIFE TO OTHERS.
YOU HAVE THE POWER OF DEATH.
THIS DEATH IS PURIFICATION FOR NO
SPIRIT DIES.
EVERYONE IS IN KRISHNA

CONSCIOUSNESS.
SLOW DOWN, YOUR MIND IS SO MUCH FASTER THAN THE WORLD MOVES.
PREMA IS THE HIGHEST VIBRATION OF LOVE IN THE UNIVERSE AND IT COMES FROM GOD OR KRISHNA.
MOST OF LIFE IS IN THE PRAKRUTI OR MOTHER NATURE AND ARE THE EFFECT OF MOTHER NATURE.
YOU ARE IN THE DAIVI PRAKRITI WHICH IS THE INVISIBLE FORCE FIELD OF LOVE AROUND THE PSI BEING.
THIS IS CALLED AFFINITY AND IT CAN BE THE STRONG LIKE OR DISLIKE OF SOMETHING, SOMEONE, AND SOME PLACE.
THE PSI BEING CAN ALSO RISE TO THE PARA-PRAKRITI WHICH IS A STRONGER VIBRATION OF LOVE IN THE UNIVERSE.
THIS LOVE IS BEST FOR WORKING ON SELF.

NOW THE SOUL IS PERFECT SO IT IS ONLY THE MIND AND BODY THAT YOU CAN IMPROVE, AND YOU CAN IMPROVE THE MINDS AND BODIES OF OTHERS AT YOUR DISCRETION.

A PSI BIENG IS NOT OBLIGATED TO HELP ANYONE BUT HE IS OBLIGATED TO HELP HIMSELF.

HE OR SHE MUST KEEP THEIR ETHICS IN AND DO THEIR PSI EXERCISES.

DO NOT FEEL BAD TO DELIVER A BLOW FOR A GOOD CAUSE.

BUT THIS IS ONLY FOR SELF DEFENSE OR OFFENSE.

NO ONE DIES EVERYONE IS TRANSFORMED AND PURIFIED.

THE PARAMATMA IS IN EVERYTHING AND EVERYONE.

SOUL IS THE VITAL SOURCE, INDESTRUCTIBLE AND PURER THAN THE PUREST.

YOU HAVE THE SUPERIOR LOVE, THE
LOVE OF THE ETERNAL LIVING GOD.
OUT OF YOUR PHYSICAL BODY {MATTER},
INTO YOUR PSYCHIC {MASS} BODY.
DAMN! YOU WOULD THINK THIS EASY.
BUT I STRUGGLE FOR ALL THE IDEAS THE
WONDERFUL THEORIES THAT HELP
MANKIND, ALL OF THEM ALMOST
DISAPPEAR WHEN I LOOK AT A BANK
SHEET OF PAPER.
ONE GUY ASKED ME 'WHERE DO ALL
THOSE WORDS COME FROM?'
'OUT OF THE ISSUES OF THE HEART?'
IF YOU GOT ALL YOU NEED, WHERE IS
THE DAILY DESIRE TO LIVE?
IN THIS WORLD, EVERYTHING IS A SIN,
WHERE DOES THE GOOD MAN GO?
I DO WAIT ON PSI PRECEPTS!
THIS IS A HAPPY SPACE.
THIS IS A HAPPY TIME.
THIS IS A HAPPY PLACE.

THIS IS A HAPPY ENERGY.

THIS IS A HAPPY MATTER.

THIS IS A HAPPY SOUL!

I AM NOT MY PHYSICAL ENVIRONMENT OR THE PROCESSES OF LIFE.

I AM NOT MY PHYSICAL BODY OR SKIN OR ELECTROMAGNETIC BODY.

I AM NOT MY PSYCHIC BODY OR ASTRAL BODY OR LIGHT BODY.

I AM NOT MY FALSE EGO THAT THINKS I AM A ONLY A PSYCHIC OR PHYSICAL BODY.

I AM A SPIRITUAL BEING {SP}, MY PSI BODY IS FULL OF ETERNAL BLISS, KNOWLEDGE AND LIFE {ENERGY}.

I AM A PSI BEING.

I WAIT ON PSI PRECEPTS FROM THE MASTERMIND.

THE MASTERMIND WITHIN AND WITHOUT.

THE MASTERMIND CALLED WITHIN, MY

SUPERMIND.

MY SUPERMIND AND MASTERMIND ARE ONE.

WE WERE THE DEMONS THAT CREATED HELL!

BUT HELL IS 'OTHER PEOPLE' THE LACK OF LOVE OR THE STRAIN OF LOVE.

HELL IS OTHER PEOPLE, THE HELL THEY BUILD FROM ABOVE.

HELL IS ANOTHER PERSON'S MIND, BELIEVING IN LIES OR ILLUSIONS.

HELL IS IN YOUR MIND'S EYE AND ANOTHER PERSON'S DELUSION.

TIME WAITS ON NO MAN, WOMAN, OR CHILD.

YOU ARE THE BEST IN THE WORLD.

THE WORLD NEEDS YOU; YOU DON'T NEED THE WORLD!

YOUR PSI BODY IS YOUR TRANSCENDENTAL BODY.

THE BODY OF KRISHNA.

AUM TAT SAT!

THIS IS A NEGATIVE WORLD. DON'T LET IT SELL YOU ON EVERY NEGATIVE THING. NOW WE ARE MASTERS OF OUR SENSES, MIND, BODY AND IMMEDIATE ENVIRONMENT.

WE ARE ENOUGH. ALL WE HAVE IS ENOUGH. KRISHNA IS ENOUGH.

YOU ARE HIGHER MATHEMATICS:

1+ 1 = 1 and 1 - 1 also EQUALS ONE.

THIS IS THE SPIRITUAL WORLD.

YOU ARE A PSI BEING!

The Last Day or the First Day of the last page.

Here is the end of THE PSI TECH 2023.

FINISHED in JUNE 17 days before my 70th birthday.

This is how it ends.

SO WHAT IS THE ENDING OF SOMETHING THAT IS ETERNAL?

WHY OF COURSE IT CAN HAVE NO END EXCEPT WITHIN ITSELF.

START-CHANGE-STOP.

THAT IS EASY NOW AND CAN BE APPLIED TO ANY ACTION, EFFORT, E-MOTION, OR E-GO.

YOU CANNOT DIVORCED THE BODY FROM THE EARTH NO MORE THAN YOU CAN DIVORCED THE EARTH FROM THE UNIVERSE.

THAT WHICH IS FIRST CAUSE CANNOT BE DETERMINED UNLESS THE FIRST CAUSE HAS 'NO CAUSE' TO EXIST AND THIS WOULD DENOTE ETERNITY OR AN ETERNAL THING, PLACE OR BEING{ness}.

THEREFORE, WE CAN IMPOSE ON US AND OUR UNIVERSE A FIRST CAUSE WHICH WOULD BE THE CAUSE OF ALL CAUSES THAT NEED NO CAUSE TO EXIST AND THAT WOULD BE THE SUPREME BEING{ness}, GOD, OR KRISHNA.

THOUGH, KRISHNA IS NOT GOD BUT THE ETERNAL SOURCE OF ALL GODS.

AND THE ONLY GOD OF AFRICA SAID: "KRISHNA IS THE NAME OF GOD GIVEN TO ALL MANKIND."

AND KRISHNA SAID:"YOU DO NOT HAVE TO WORSHIP ME OR KNOW MY NAME, BUT THOSE WHO DO ARE HELPED BY ME AND PROTECTED BY ME."

KRISHNA IS BEYOND WORDS, SOUNDS, ALL VIBRATIONS, THE PHYSICAL AND PSYCHIC WORLDS, AND THE SPIRITUAL AND PSI WORLD.

KRISHNA IS TRANSCENDENTAL AND SO ARE HIS PSI BEINGS.

A SPIRITUAL PERSON MAY BE A GOOD PERSON AND NOT KNOW ANYTHING ABOUT THE TRANSCENDENCE.

AS FAR AS KRISHNA AND THE PSI BEINGS ARE CONCERNED,

' IF IT IS NOT KRISHNA, IT IS ILLUSION.'

BUT 'NOT KRISHNA IS AN IMPOSSIBILITY
UNLESS KRISHNA ALLOWS IT.
WHEN R > (o) ONE IS IN A DEVOLUTION
STATE OF MIND AND COULD GET WORSE.
WHEN R = (o) ONE IS IN A EVOLUTION
STATE OF MIND AND CAN GET BETTER,
BUT NORMALLY GET WORSE.
WHEN R > (o) ONE IS IN A INVOLUTION
STATE OF MIND CAN GET BETTER
PROVIDED HE OR SHE 'PULLS HIMSELF UP
BY HIS OWN BOOTSTRAPS.'
WHEN R < (o), and R is = (o), AND R > (o)
ONE IS IN A TRANSCENDENTAL STATE OF
MIND.
TK.
THE WHOLE PURPOSE OF THE PSI TECH
IS
TRANSCENDENTAL CONSCIOUSNESS and
TRANSCENDENTAL MEDITATION OR
LIVING.

OUTRO

NO WORDS OR SENTENCES SHOULD BE ADDED TO THIS BOOK AND THIS BOOK SHOULD NOT BE ALTERED IN ANY WAY.

THIS BOOK HAS ELEVATED YOU TO THE ETERNAL ABODE OF GOD.

THE ETERNAL ENERGY WE ARE ALL PART AND PARCEL OF BE WE THEIST, ATHEIST OR AGNOSTIC OR SOME OTHER WEIRD TITLE IF ONE IS NEEDED.

THIS ETERNAL ENERGY MAKES YOU IN

ESSENCE PSI BEINGS FOR PSI BEINGS ARE ANTI-MATER BEINGS OF WHICH ALL MANKIND CONSIST OF KNOWINGLY OR NOT.

THIS IS NOT A SUBJECT OF AGREEMENT, BUT OF FACT.

AS SURELY AS THE EARTH ROTATES AROUND THE SUN.

THERE IS NO HIGHER KNOWLEDGE THAN THIS FOR THIS IS TRANSCENDENTAL KNOWLEDGE OF THE HIGHEST ORDER.

KRISHNA SAT IN THE VOID AND THE NOTHINGNESS AND THE DARKNESS AND THE BLACKNESS WHICH WAS HIS OWN FOR NOTHING ELSE WAS YET CREATED AND THE LIVING ENTITIES DANCED LIKE PEARL ON A STRING UNDER THE REIGN OF THEE.

YOUR ETERNAL SELF SEES AND RECORDS ALL THAT YOU DO, LIKE YOU ARE THE STAR OF YOUR OWN INNER MOVIE THE

WORLD MAY NOT AGREE.

BUT YOU NOW KNOW TRUTH FROM FANTASY OR SO YOU THINK!

YES, YOU DO. THE EXTERNAL AND THE INTERNAL WORLDS.

THIS WORLD IS BUILD TO CONFUSE YOU!

STAY IN YOUR OWN WORLD!

HARE KRISHNA, HARE KRISHNA!

KRISHNA! KRISHNA! HARE! HARE!

HARE RAMA! HARE RAMA! RAMA! RAMA!

HARE! HARE!

THIS IS THE BEST WAY, THE RIGHT WAY, THE EASIEST WAY.

YOU CREATE NO KARMA.

I AM SO HAPPY YOU CAME MY WAY AND I HAD THIS CHANCE TO REACH YOU, TO TEACH YOU, TO SHOW YOU THE WAY.

AND IT ALL COMES DOWN TO TWO THINGS:

MEDITATION AND TK!

THERE IS NOTHING MORE TO SAY.

101 WORDS!

WE HAVE NO NEED TO FEAR DEAD PEOPLE, FOR DEAD PEOPLE SEE US AS DEAD TOO.

THERE IS NO 'OUR GOD' OR 'OH MY GOD!'

GOD DOESN'T BELONG TO US; WE BELONG TO GOD!

THE PSI BEING ALSO EXIST IN THE NON-EXISTING MIND OF SATAN.

SO. THERE IS A GOD OF GRACE AND A GOD OF MERCY. A GOD OF GOODNESS AND A GOD OF PASSION.

AND A GOD OF IGNORANCE.

THERE ARE THE HOLY BOOKS, THE SPIRITUAL BOOKS, THE SECRET KNOWLEDGE, AND THE TRANSCENDENCE.

THIS IS THE TRANSCENDENCE.

AND YOU ARE THE PSI BEING.

THE MOST POWERFUL BEING(ness) IN ANY UNIVERSE.

KRISHNA SAID SO.

AUM TAT SAT!
I FOCUS ON MY INNER SELF. MY ETERNAL
SELF OF JOY AND KNOWLEDGE!
HOW PAIN TRIES TO DRIVE ME TO ANGER,
TO HATE, TO ATHEISM!
HOW CAN GOD LET ME SUFFER? AFTER
ALL I GAVE MY LIFE TO HIM!
THE PAIN AND THE JOY, THE KNOWN AND
THE UNKNOWN!
PSI BEINGS ARE NOT GUILTY OF
ANYTHING.
THEY HAVE FORGAVE THEMSELVES AND
OTHERS IF NECESSARY.
ALL IS A CLEAN SLATE. ALL IS A NEW DAY!
A NEW WAY!
I CAN DO SOMETHING ABOUT
EVERYTHING!
I AM A POSITIVE LIGHT IN THIS DULL,
DEAD, DARK WORLD!

I BRING THE LOVE, THE JOY, THE WISDOM, THE FUN AND THE PLAY.
WHERE WILL MY MIND GO? BACK TO THE CRAP IN MY HEAD OF WHAT HAPPENED TO ME THAT MADE ME THIS WAY?
OR DO I GO TO KRISHNA, MY ETERNAL SELF OF JOY, KNOWLEDGE, AND BLISS?
HARE KRISHNA, HARE KRISHNA, KRISHNA, KRISHNA, HARE! HARE!
HARE RAMA, HARE RAMA, RAMA, RAMA, HARE! HARE!
I AM FREE AND DECIDING WHAT TO BE TODAY!
WHAT AM I CONSIDERING?
YOU CAN FLOW ENERGY TO YOU.
YOU CAN FLOW ENERGY FOR YOU/
YOU CAN BE IN YOUR BODY {not recommended}, or
OUT OF THE BODY NEAR THE HEAD {recommended}.
YOU POSSESS ALL THE QUALITIES OF THE

UNIVERSE, SPIRITUALLY YOU LACK
NOTHING.
ALL THAT IS GOOD IS WITHIN YOU AND
YOU HAVE DIVINE ATTRIBUTES.
YOU CREATE YOUR OWN SPACE, TIME,
ENERGY, MASS, AND MIND.
ALL IS IN MY MIND.
TK!
YOU CARRY WHAT YOU WANT AND DESIRE
AND SUSTAIN WHAT YOU ALREADY HAVE.
YOU CAN BE, DO, AND HAVE!
LET NOT THE WORLD PUNISH YOU.
YOU ARE AS YOU ARE!
DON'T FEAR.
DON'T HESITATE!
DON'T WORRY.
KIRSHNA IS ALL.
TK.
BEFORE ABRAHAM I AM?
BEFORE "I AM"!
"I AM!"

ALL RELIGIONS CAME AFTER ME!
ALL PHILOSOPHIES ARE OF ME!
I WAS NO BORN NOR CAN I DIE.
AND I AM YOUR BES OF FRIENDS
TRUST IN ME LIKE I TRUST IN YOU!
WHAT CAN I KNOW THAT YOU DON'T
KNOW?
DON'T BE AFRAID TO ASK FOR HELP!
EVEN THE BEST WAS NOT THE BEST AT
FRIST.
YOU ARE A PSYCHIC WARRIOR!
YOU DO HAVE PSYCHIC POWERS!
EXTENDED SENSORY PERCEPTION!
YOU SEE WHAT OTHERS DO NOT SEE.
YOU SEE WHAT OTHERS ONLY DREAM OF.
YOU ARE THE ACTOR IN YOUR DREAM!
DREAM NO MORE!
YOU DON'T NEED TO KNOW EVERYTHING
TO BE HAPPY AND FREE,
JUST CHANT AND THINK OF ME!
AUM TAT SAT!

YOU CANNOT SAVE THE WORLD.

PEOPLE ARE BOUND BY EMOTION, NOT RATIONALITY!

A PSI BEING IS NOT CRAZY.

HE IS A POTENTIAL SPARK OF LIGHT.

HE IS SO LIGHT, HE CAN VIEW HIS BODY AND ENVIRONMENT FROM A DISTANCE AND STILL BE IN CONTROL OF IT.

THIS IS A GOOD DAY TO BE A GOOD EFFECT OF THE DAY!

I FEEL THE WIND, I AM THE WIND, THE WIND IS ME!

WHAT A GREAT EXPERIENCE!

MY BODY STARTS TO HEAL ITSELF.

I JUST DO WHAT IS RECOMMENDED!

DO YOUR OWN WORK, THE JOURNEY IS NOT THE SAME FOR EVERYONE.

DON'T WORRY.

DON'T HESITATE.

DON'T FEAR!

YOUR LIFE IS AN ART. AN MEDITATION.

THIS IS THE BEST SOLUTION.
THE BEST MOTION.
HOW EASY IT IS TO DO NOTHING!
I HAVE NO OBLIGATIONS.
NO ONE I MUST ATTEND TO BUT MYSELF.
I RELAX AND SMOKE AWHILE.
I ALWAYS GIVE MY BEST. I REST.
I FEEL KRISHNA DOING IT FOR ME I
ALMOST PROTEST!
MY CHOICE TODAY GOOD OR BAD, HAPPY
OR SAD, MAD OR HAD!
I AM NEUTRAL. TRANSCENDENTAL.
MOTIVATIONAL.
I GIVE LIFE.
HAPPINESS IS WHAT I THINK IT IS.
JOY IS THE LAUGHTER OF THE UNIVERSE.
KNOWLEDGE IS WHAT I SEEK
EVERYWHERE.
I AM RIGHT AND I KNOW IT!
I BREATHE IN DEEPLY, EXHALE
COMPLETELY!

I AM THE LIFE SOURCE OF EVERYTHING.

A STATIC AND A KINETIC I GET THINGS DONE!

MY MIND, BODY, AND ENVIRONMENT ADJUSTING TO KRISHNA.

I AM IN MOTION AND TRANSCENDENT!

I LOCALIZE ABOVE MY PHYSICAL HEAD.

ALL YOGA LEAD TO THIS.

ALL SPIRITUAL EXERCISES LEAD TO THIS.]

ALL OF LIFE LEAD TO THIS!

TRANSCENDENTAL CONSCIOUSNESS!

THOUGH IT IS ONE AND IT IS THE SAME, IT IS

DIFFERENT FOR EVERYONE!

TK!

DON'T SPEND YOUR LIFE TRYING TO FIND THE REASON

WHEN YOU ARE THE SOLUTION TO YOUR PROBLEM.

AND NOW YOUR BIGGEST PROBLEM OF

ALL IS ONLY FORGETTING WHO YOU ARE!

AUM TAT SAT!

THIS IS WHERE IT'S AT!

SAMADHI!

THE LIVING UNIVERSAL SOUL IS AN ETERNAL PART OF ME IN BLISS AND LIFE AND GREAT KNOWLEDGE.

I AM A PSI BEING!

END OF PSI TECH

About the Author

Being raised a Christian all my life I always looked for ways to improve my 'walk with God' and found that the more I did what the church and bible told me to do the sicker and more tired I became of life and ended up hating myself and loving others more than myself.

I never was good enough for anything, and being a 'sinner' I did the things that sinners did and then repented to find it all a crazy loop of save and back slide and save and back slide and a horrible inner battle where sometimes the good guy won, but most the time the bad guy or (carnal man} would will out!

I hated myself and tried to be the best Christian alive.

This led me to the Navy and being an Electronics Technique and also as a Missionary for The Way Ministries.

It was then that my eyes began to open and I could see I was merely a pawn in the hands of the church who were only using my guilt and 'sinful nature to be good' as a way of making money off of us and in return we 'suffered for Christ' as we worked for The Way Ministries.

Everything I did was 'Christian' in some way or another or the complete opposite where I would go against almost every law of the Old and New Testaments.

With my grandmother dressing me up in girl clothes when I was a little kid, it was easy for me to think of myself as being 'gay' but found I hated 'gays' God hated 'gays'

and this made me hate myself more and hate gays even more.
So I punished people for being gay and later on changed my heart and accepted them into society as an alternative way of sex.
Being 'gay' for the most part was acting like a 'girl' for the most part and I despised that also.
And the macho gays just had something about them that made them well 'queer' and not in a bad way.
I found that I did not really like sex at all but was a means of release and feeling good.
I bring this all up for I want people to see how sexual abuse can critically harm a child and give him a self-image that is not only negative but is not his own.
I always was a straight man but used sex to make money and for dominance.
I love to torture men in a fun way and

which is not really harmful if the person is consenting.

I got married but didn't have any children. This was a good thing for I soon started having problems with the tumor on my head that I got when my dad supposedly, accidentally, dropped me on my head as a little baby when you still had the 'soft spot' in your head. If this was a problem with me I certainly did not want to pass it on to any of my children.

The marriage did not go well anyway so I got divorced and decided to stay single the rest of my life.

With no family so to speak to take care of, I began to search for the true meaning of love and life which would include the family or not.

It was the trauma I suffered as a child that helped me to discover this new science though it is as old as Africa itself.

Not only do we have the power to heal ourselves, we can also change any negative into a positive and eliminate or lessen the pain of what is lost.

Nothing is lost in Spirit for God nor His Eternal Energy can be divided.

This is another reason METAPSYCHICS WORKS, we are dealing with a higher reality making METAPSYCHICS something that you DO, NOT believe.

METAPSYCHICS is a gift of God to you, the individual, and to mankind {every man, woman, and child}.

I am proud to be the one who helped create it for you.

After years and years of being sick and disabled, and trying everything under the sun to help me relieve some of the sometimes excruciating pain and suffering I had from being born early and other negative incidents in my life, I finally found something that show me my reason and purpose for living and how I can use it to help others live a better life.

It was only after I got tangible results from METAPSYCHICS I presented it to th

What I found will astonish you for it is over 50 years of research into LIFE, and the REASON FOR LIVING.

In short, METAPSYCHICS is more a DISCOVERY than a CREATION of mine.

I only ORGANIZED and CODIFIED it and brought this ancient science of God, Soul, and Soul Travel into a modern science and spiritual science you can KNOW, UNDERSTAND, and USE for the benefit of yourself as well as others.

The Science and Spiritual Science of MetaPsychics is the work of one man and is a gift to Man and Mankind.

{At a nominal fee.}

The cost of the publications is only to keep

METAPSYCHICS FREE TO THE PUBLIC and the website up-to-date.

Discover more about yourself, how you can heal yourself and heal others, and use your PSI POWERS to create the person and universe you always wanted to be and live in.

www.ingramcontent.com/pod-product-compliance
Lightning Source LLC
LaVergne TN
LVHW090358210926
848480LV00002B/384
* 9 7 9 8 2 3 0 9 0 7 0 1 5 *